Women Pioneers for the Environment

NORTHEASTERN UNIVERSITY 1898–1998

Women Pioneers for the Environment

MARY JOY BRETON

Northeastern University Press
Boston

Northeastern University Press

Library of Congress Cataloging-in-Publication Data
Breton, Mary Joy, 1924–
    Women pioneers for the environment / Mary Joy
Breton.
        p.   cm.
    Includes bibliographical references and index.
    ISBN 1-55553-365-5 (cloth : alk. paper)
    ISBN 1-55553-426-0 (pbk : alk. paper)
        1. Women environmentalists—Biography.   2. Green
movement—History—19th century.   3. Green movement—
History—20th century.
I. Title.
GE55.B74   1998
363.7′0092′2—dc21
[B]   98-16439

Designed by Janis Owens

Composed in Janson by Coghill Composition, Richmond,
Virginia. Printed and bound by Thomson-Shore, Inc.,
Dexter, Michigan. The paper is Glatfelter Supple Opaque
Recycled, an acid-free sheet.

MANUFACTURED IN THE UNITED STATES OF AMERICA
02   01   00   5   4   3   2

TO

Hazel Wolf

ILLUSTRATIONS

Hazel Wolf had been pushing me to write this book since the early 1980s. At her request, I had submitted an article about women in conservation to a publication she edited, *Outdoors West*. After spending a winter's worth of weekends doing research, I sent her all the material I had collected beyond what I could put into the short article. With my permission, she began using the information in talks around the country. In the spring of 1995, a few months before I left my position with the National Audubon Society after sixteen years, Scott Brassart, an editor for Northeastern University Press, saw an excerpt from one of Hazel's talks and approached her about doing a book. She referred him to me.

Women's concern for the health of the planet dates back centuries. For example, Hildegard of Bingen (1098–1179), a German nun, writer, artist, and mystic, forewarned about the ecological peril now facing the earth.[1] From the beginning of the more recent environmental movement, women have been the force driving grassroots activism. This book does not pretend to be exhaustive; it is, rather, merely a sampling of women environmental activists from around the world. My most difficult task has been deciding who, from among the army of women my research uncovered, to include in the book. The realities of length, along with the need for geographical and ethnic balance, forced me to keep narrowing my list, a painful process.

For example, three women whose pioneering behavioral field studies with apes—the late Dian Fossey with gorillas, Jane Goodall with chimpanzees, and Biruté Galdikas with orangutans—have led to significant breakthroughs were on my original list. As Ashley Montagu says of them in *The Natural Superiority of Women*, "Where men had gone into the wild armed with guns, the women went in armed with nothing more than goodwill, tact, and delicacy, and succeeded in making many new discoveries where generations of men had failed."[2] In taking Fossey, Goodall, and Galdikas off my list, I was comforted by knowing that Sy Montgomery has given a comprehensive account of their achievements in her wonderful book, *Walking with the Great Apes*.[3]

Of necessity my pieces about individual women are short, scarcely doing justice to the scope of their work. I hope, however, that the stories will stimulate readers to further explorations.

Please note that, unless otherwise designated, all quotations from the principals are from personal communications with me.

ACKNOWLEDGMENTS

First and foremost, I thank Hazel Wolf for never giving up for fifteen years in urging me to write this book. I thank her as well for steering my editor, Scott Brassart, in my direction.

My daughter, Denise, and her husband, Chris Largent, both published writers, not only gave me loving encouragement and support but also provided valuable editorial suggestions.

Jean Kaufman, a reference librarian at the Wilmington (Delaware) Public Library, became a buddy, helping me dig out needed information. No request daunted her. I also appreciate the help of her colleague, Ben Prestianni.

Other reference librarians and archivists assisting me include Margery N. Sly, Smith College Library; Charles Longley, Curator of Microtexts and Newspapers, Boston Public Library; Liz Andrews, Reference Archivist, MIT Library, and Elizabeth Kaplan, Assistant Archivist, Institute Archives and Special Collections; Michael Yates, MIT Museum Collection; William T. Milhomme, Reference Supervisor, Commonwealth of Massachusetts Archives, Boston; and Maureen Heher at the Beinecke Rare Books and Manuscripts Library, Yale University.

I would especially like to acknowledge that in addition to numerous other sources, I derived much of the material in chapter 2 from Robert Clarke's and Carolyn Hunt's biographies of Ellen Swallow (Richards).

My thanks also to Kathleen McGinty, Chairwoman of the President's Council on Environmental Quality; Peter A. A. Berle, host of WAMC/Northeast Public Radio's *Environment Show*, former Commissioner, New York State Department of Environmental Conservation, and former President and CEO of the National Audubon Society; Susan Murcott, Research Environmental Engineer at MIT, and film producer Carma Hinton; Frank Graham, author and *Audubon* Field Editor; Wallace Dayton, former chairman of the Nature Conservancy; Elaine O'Sullivan, Norman Brunswig, Walt Pomeroy, Don Arnosti, Larry Thompson, Bucky Dennerlein, and Connie Isbell of the National Audubon Society staff; Bonnie Kreps of the Craighead Environmental Research Institute; Taher Husain at the Rachel Carson Council; Karen Peterson at the Goldman Foundation; Sara Urdahl of Sleeping Lady Lodge; Denise Brady, Port Washington, New York, Public Library; Deborah Boyd, World Wildlife Fund; President's Office staff, the Nature Conservancy; LaVonda Walton at the U.S. Fish and Wildlife Service; Donna Carroll of the Gwich'in Steering Committee; Betty Ball, Mendocino Environmental Center; Anne Champagne, Valhalla Wilderness Society; Alison Kahn, International Society for Ecology and Culture/Ladakh Project; Mary Kay Feezer, Moose, Wyoming; Chisato Murakami, Environment Partnership Office, Tokyo; Alan Tillotson of the Chrysalis Center, Wilmington, Delaware; Marilyn Fike, Administrator, Bullitt Foundation; Nancy Mbura Nyaga of the Green Belt Movement; and Robin Mitchell and Charles Pederson at the Florida Defenders of the Environment.

Special thanks to Dr. Tatyana Artyomkina of Ryazan, Russia; Professor Maria Gumińska of Kraków, Poland; Joyce Andreae of England; and Anneke and Yfke Jepma and David Andreae of the Netherlands for their helpful suggestions regarding marketing this book in Europe.

Tony Granados of Granados Associates, the author Joan Ohanneson, Lori Russell at Waldenbooks, Amy Forster at ISAR in Washington, D.C., Jeannine Baden, Asha Bhaya, and Laura Shelton also have my gratitude for their help.

Finally, I appreciate the free rein and flexibility afforded by my editors at Northeastern University Press, Scott Brassart and John Weingartner.

In England a woman chains herself to a bulldozer and is jailed for trying to prevent destruction of a unique habitat.

In Russia a woman scientist—though threatened by the KGB—goes public with information on health and environmental hazards she believes citizens have a right to know about.

In Poland a woman biochemist effects the shutdown of a major polluter, the country's largest aluminum plant.

In Kenya a woman is beaten and imprisoned when she extends her environmental activism to human rights issues.

In China a woman journalist is imprisoned for opposing an environmentally destructive dam on the Yangtze River.

In Michigan a woman is ostracized by her friends and neighbors for opposing construction of an unsafe nuclear plant.

In the Pacific Northwest a woman is crippled by a car bomb, planted because of her efforts to protect ancient forests.

In 1892 a world-renowned woman sanitary scientist and expert consultant on water quality—the first woman to attend MIT—introduces the word "ecology" in the United States.

❧ What motivates such women? How do they persevere against the odds? What qualities or characteristics do they share?

Do women have a particular sensitivity to Earth? Do they bring special insights, value systems, and worldviews to environmental issues? What is different about women's approach to resolving ethical, ecological, and economic dilemmas?

What do we lose by excluding women's unique perspective from decision making—and how do women overcome this obstacle?

How do women get heard? How do they transform entrenched cultures? How are they effecting new relations with the planet? What are women doing to prepare the next generation for the inevitable ecological revolution?

The answers may lie in the lives of nineteenth- and twentieth-century women who stepped out of their traditional, subservient roles and created profound changes that help ensure the fundamental right of all living things to a healthy world.

Never underestimate the power of groups of committed citizens to change the world. In fact, it is the only thing that ever has.

MARGARET MEAD
"Our Open-Ended Future"

Tree Huggers and Tree Planters

MARTYR AMRITA DEVI
Inspires the Chipko Movement

As the maharajah's axemen approached the first tree marked for felling in the heavily wooded district of Rajasthan, in India's Himalayan foothills, Amrita Devi wrapped herself around its trunk. The inhabitants, adherents of the Bishbios religious sect, held trees as sacred. Each child, for example, had a special tree to talk to and hug. But the maharajah of Jodhpur, wishing to build himself a new palace, had dispatched a crew to chop down trees to fire his lime kiln. The axemen ignored Amrita Devi's pleas to spare the forest. As she clung to the tree, crying "A chopped head is cheaper than a felled tree!" the axe came down. After she had crumpled to the ground, her three daughters each in turn took her place defending the trees. All were killed. Then, persons from forty-nine surrounding communities responded to the villagers' call for help.

Facing a major confrontation, the axemen warned the villagers that resistance would mean death for them also. They continued to hug the trees, refusing to yield. By day's end over three hundred and fifty women and men had been slaughtered.

The maharajah, on learning why the tree cutting had been progressing so slowly, had a change of heart. He abandoned the palace-building project, ordered a halt to the cutting of trees, and went to the scene of the violence to apologize to the villagers. He promised that never again would their trees be cut.[1]

Thus did the Chipko movement of India begin, three hundred years ago. In both Garhwali and Hindi, *chipko* means "to embrace."

Inhabitants of Western countries like to believe they originated the ecological movement. More likely, these villagers, though without formal education, gave it birth. The concept of all life forms as interconnected, sacred, and part of the whole—and therefore to be respected—is fundamental in their cultural heritage.

For centuries Indian women managed the forests of the steep Himalayan foothills as a communal resource. The trees in this important Asian watershed held soil in place,[2] put oxygen in the air, and stored and purified the monsoonal rains, slowly releasing water into catchments during the dry season. The forest also provided the region's inhabitants with fuelwood, leaves for animal fodder, shelter materials, medicinal herbs, mushrooms, fruit, honey, and nuts, as well as raw materials for local crafts.

To villagers, the hillside forests were a matter of life and death. Without trees, the naked mountain could come crashing down in mudslides or unleash floods that could carry houses away.

Until India's major road-building campaign began in the 1960s, Himalayan forests were difficult for commercial loggers to reach. But after the state began carving roads into the hillsides, the Indian Forest Department, viewing trees as a cash resource, let contracts to commercial timber interests. Ignorant of how the Himalayan watershed ecosystem functioned, the loggers soon clear-cut large mountainside areas, inviting environmental and economic disaster.

In the 1960s, when small forest industries burgeoned in the hilly region, conflicts arose in the villages between men, who wanted cash from commercial forest products, and women, who wanted to conserve the forests as local life-support systems. For village women, food production began with the forest, and the disappearance of the trees was an issue of survival as well as of human rights. Although they suffered most from environmental degradation, women had been excluded from the decision making that had allowed the destruction.

Alarmed by what they saw happening, and recalling the story of Amrita Devi, Himalayan village women resurrected the Chipko nonviolent resistance movement in an effort to save the remaining forests. Their only weapons were fearlessness and lack of greed. But could impoverished villagers subdue wealthy logging interests and corrupted agency officials?

Bimala Behn and Sarala Behn are credited with establishing this century's Chipko movement as a women's crusade, while Mira Behn provided its philosophical and ecological underpinnings.[3] Mira Behn, an English woman and a close disciple of Mahatma Gandhi, had moved to the Himalayan Garhwal region in the late 1940s to study the ecology of the Himalayan forests and the link between deforestation and cyclic water crises. One day she watched flood waters spill down the Himalayan mountainside from the Ganga catchment area, uprooting bushes and trees and washing away topsoil, cattle, and even humans clinging to the fragments of huts. Searching on horseback for the disaster's cause higher up the mountain, she discovered extensive deforestation—steep slopes stripped bare, then gouged by landslides.

Mira Behn established a modest center, or ashram, in the village of Rishikesh in Bhilangana, a Himalayan valley, where she focused on the region's forest problem. She learned from older local residents that catchment areas, once covered with indigenous broad-leafed trees, had been replanted with more commercially valuable pine and eucalyptus. The pines, planted by outside forest contractors, robbed the inhabitants of the economic and survival features of the broad-leafed trees, which, among other things, provided animal fodder. The eucalyptus trees had high water uptake and contributed almost nothing to the formation of humus. Mira Behn concluded that unless the Ganga catchment area was replanted with broad-leafed trees drought and floods would worsen. As early as 1949, she had written a paper entitled "There Is Something Wrong in the Himalayas."

Sarala Behn, also a close disciple of Gandhi, had set up an education center for hill women in Garhwal, where she encouraged them to regard themselves not as beasts of burden but as powerful "goddesses of wealth." The women provided almost all of the region's labor, producing food and raising cattle. In 1975, on her seventy-fifth birthday, Chipko activists honored Sarala Behn with a commemorative newspaper column, calling her the daughter of the Himalayas and the mother of social activism in the region. In 1977 all the activists in the hills congregated at her ashram to consolidate their resistance work.

The Chipko movement, comprising mostly women, has no formal structure or headquarters, no board of directors, titles, or officers. As a cooperative, nonviolent resistance movement to preserve basic rights, its staying power resides in its community-based control. It spreads by word of mouth and by the efforts of local, largely rural women connected to each other horizontally rather than vertically. It eschews hierarchy for decision making by consensus. For these reasons information about individual Chipko leaders is scant. But thousands of courageous women of all ages have joined the protests against logging, often risking their lives. Some of their stories have been recorded. Here are a few of them, as recounted by Vandana Shiva in her book *Staying Alive.*[4]

Kangad, a hamlet of two hundred families in a Himalayan valley six thousand feet above sea level, had suffered forest degradation. Only a small patch of trees remained. Women had to walk great distances each day for water, fuel, and animal fodder, while many of the village men worked for the Forest Department's felling operations. When the last patch of woods was earmarked for destruction, the women spread the word and mobilized to save it, defying their men. After a four-month resistance campaign, and with the help of the experienced Chipko activist Bimala Behn, the women won. They then embarked on a project to re-generate their degraded forests. The first step was to raise money to hire a forest guard. When he turned out to be corrupt—allowing selected persons to extract more than their share of fuelwood—the women dis-missed him and guarded the forest themselves.

During the 1960s and early 1970s, the issue of logging Himalayan forests kept simmering, and women villagers stepped up their protests. One Chipko leader, Hima Devi, traveled from village to village, speaking at demonstrations and at protests against forest auctions. In 1968, at Ti-lari, Chipko activists pledged to save the forests, recalling a historic event that had occurred there. In 1930 the British had tried to take the forests away from the local people and claim territorial rights. When a large protest demonstration took place, violence erupted and several persons were shot.

A turning point came in 1973. Three hundred ash trees in the Gopeshwar forest near the small village of Mandal had been auctioned by the state forest department to the Simon Company—a sporting goods manufacturer—to make tennis rackets. At the same time, the government refused the villagers' request to fell twelve ash trees for making needed agricultural implements such as plows. Inhabitants of Mandal marched to the forest, beating drums and insisting they would hug the trees to prevent their being cut. The loggers withdrew. Later, the villagers learned that the sporting goods manufacturer had contracted for trees in another forest some distance away. They walked to the site, set up a round-the-clock vigil, and began hugging the trees and chanting Chipko slogans until, about a week later, the contractor withdrew. The forest department canceled the Simon Company's logging permit and assigned the trees to the local villagers.

Chipko activists then shifted their focus to the Alakananda River Valley in the Himalayas. In 1970 a severe flood had inundated several villages; one had been totally swept away because a landslide had blocked the river. Almost two hundred people died. The local women of Reni linked the catastrophe with the felling of trees in a catchment area. So in 1973, when a young woman grazing her cows noticed men carrying axes and heading for the nearby forest, she summoned her neighbors with a whistle. The women surrounded the axe-carrying contractors and explained to them the forest's subsistence value to the villagers and why it should be conserved. Experienced Chipko women in their fifties, who had assumed leadership of this protest, organized small-group vigilance parties to watch the axemen. This forced the government to ban commercial logging in the Alakananda catchment for ten years.

Encouraged by this victory, Chipko activists turned their attention to the hill districts of Uttar Pradesh, where in 1975 deforestation posed a threat of landslides and soil erosion to more than three hundred villages. Here, the activists hoped to achieve a total ban on logging in the region. They embarked on two long treks in 1975—one lasting seventy-five days and the other fifty days—to mobilize public opinion. As a result, during the next few years the Chipko activists saved six forested areas and compelled a fifteen-year ban on logging.

Advani, one of these areas, was auctioned to a forest contractor in October 1977, despite the ban and local opposition. The contractor planned to begin cutting during the first week of December. As tokens of their protection vow, the women fastened sacred threads around the trees. For a week women from fifteen villages guarded the forest while carrying on continuous readings from ancient texts about the forest's role in traditional Indian life. The contractor retreated, but he reappeared six weeks later with two truckloads of armed police. They planned to surround the forest to keep out the activists while the trees were cut.

Meanwhile, Chipko volunteers had already gone into the forest to talk with the laborers, who had been brought in from distant areas, educating them on forest ecology. By the time the police and the contractor arrived, each tree had three people guarding it. The contractor gave up.

Whether forest destroyers were locals or outsiders made no difference to the Chipko women. The activist Bachni Devi led a large protest against her own husband, who held a contract to cut a forest. When the forest contractor came to the scene to challenge the Chipko activists, he discovered the women carrying lighted lanterns in daylight. The women told the puzzled forester that they planned to enlighten him about forest management. He sneered, calling them "foolish women" who did not understand the value of forests in producing profits from resin and timber. In reply, the women sang back in chorus their standard verse about the forest's ability to sustain the earth with soil, water, and pure air.

The writer Brian Nelson tells the story of Kalawati Devi, a Chipko activist in the Garhwal Hills of the Himalaya region, who remembers a protest in the late 1970s involving a forest near her village of Baccher, home to a hundred families.[5] Villagers had heard about the nonviolent resistance movement. So when a private contractor began logging in a protected forest reserve Kalawati Devi led village women in confronting the loggers. The women embraced the trees and grabbed the men's axes. Kalawati Devi, never having behaved contrary to Indian tradition before and not knowing if the women's actions would end in disaster, nevertheless managed to disguise her fears.

The contractor attempted to buy off the women, offering a thousand rupees (about sixty dollars—a large amount to impoverished villagers) if they would get out of the forest. Kalawati Devi told him that the thousand rupees were not a necessity but that the trees were. For a while the logging ceased, but then it resumed. The village women's group, led by Kalawati Devi, traveled to the area's main town, Gopeshwar, to speak with the district Forest Department officials. One of them told the women that the forests belonged to the government, not to women villagers.[6] He threatened police action if the tree hugging did not cease immediately. The women were not intimidated. If the police interfered with their peaceful protest, they said, there would be even more Chipko women to greet them.

With that, a high-ranking Forest Department officer decided it was time to settle the issue for good. He and a few of his staff went to the forest, intending to lay down the law. When they arrived, they were astounded to find hundreds of women singing and chanting among the trees, many with babies strapped to them and with children playing nearby. Disarmed by the sight, the official withdrew, trying not to be noticed. No more cutting occurred.

Although most village men, favoring commercial logging for cash income, and suspicious of the wave of female activism as well, had difficulty accepting the Chipko movement, others supported the women, learning from them and helping as runners, messengers, and builders of stone soil-retention walls. A few men even assumed Chipko leadership roles. In 1952, Sunderlal Bahuguna (together with his wife, Vimla) established a Chipko information center outside a town called Tehri; it maintains a library and Chipko archive.

The movement has broadened its activities to embrace not only trees but all elements of the earth's ecological systems, from living mountains to living waters. For example, members came to realize that limestone quarrying had damaged local aquifers. To Chipko women, rocks are nature's waterworks and limestone is nature's reservoir. Yet in 1987, during a peaceful protest to halt extraction of the remaining limestone in a quarry after the lease had expired, limestone miners stoned Chipko women and children.

In 1980, the Chipko activist and educator Sarala Behn pointed out in her "Blueprint for Survival":

We must remember that the main role of the hill forests should be not to yield revenue, but to maintain a balance in the climatic conditions of the whole of northern India and the fertility of the Gangetic Plain. If we ignore their ecological importance in favour of their short-term economic utility, it will be prejudicial to the climate of northern India and will dangerously enhance the cycle of recurring and alternating floods and droughts.[7]

There is not much tree hugging anymore—the Chipko resistance movement has prevailed. After half the original Himalayan forest cover had been destroyed, the victories of the 1970s and 1980s led to bans on commercial logging in most of the region. If the Himalayan watershed were allowed to die, all of India could become a desert. The Chipko movement generated pressure for policies more sensitive to ecological requirements and inhabitants' needs.

Village women's groups, "mahila mandal dal," once again manage the forests for long-term renewability and make the rules governing the community's use of this resource. The groups engage in intensive tree planting and other watershed-restoration projects, such as terrace farming. They gather seeds from mature indigenous trees and raise their own seedlings. Children help out, learning hands-on ecology. In many of the villages these women's groups use their organizational experience to meet educational and health-care needs as well. And throughout India environmental activists have adopted the model of the Chipko movement in resisting dams, mines, and highways that would disrupt the ecosystem. The movement demonstrates people power, and it serves as a symbol for citizens in developing countries struggling for self-determination in resource management.

In 1978 the Right Livelihood Foundation gave the Chipko movement its award, often referred to as the "Alternative Nobel Prize."[8] The citation reads in part: "The Chipko movement is the result of hundreds of decentralized and locally autonomous initiatives. Its leaders are primarily women, acting to protect their means of subsistence and their communities."[9]

Since 1978 the Bishbios have been honoring Amrita Devi's seventeenth-century martyrdom with an annual fair in her village in Rajasthan. The village erected a tower on the spot believed to be the site of her murder. They also created a marble waterpool and planted over three hundred and fifty trees supplied by the government—one for each victim of the massacre.

WANGARI MAATHAI (1940–)
Recruits Army of Green Belt Foresters

Under a dictatorial regime that proclaimed it "un-African" for a woman to be anything but docile, Wangari Maathai has dared to make herself a dissident to the point of being publicly vilified, labeled "subversive" by Kenya's ruling group, ostracized, beaten unconscious, and thrown in jail. She started off innocently enough, merely wishing to help restore her country's degraded environment while providing poor rural women with better economic lives. But along the way Maathai generated resentment among the ruling political group and among husbands, including her own, who believe that women should never dispute the judgment of men.

Like so many emerging countries, Kenya has had one of its most valuable long-term assets—its forested areas—degraded by overtimbering for quick profits and to make room for export cash crops from coffee and tea plantations. Kenya retains less than 3 percent of its original forest cover. Lack of fuelwood for cooking forced changes in diet for the poor, leading to malnutrition. And each year thousands of tons of topsoil are washed down bare mountain slopes into muddy rivers and then into the Indian Ocean, destroying coastal marine life in the process.

But for two decades Wangari Maathai and a throng of women, numbering more than sixty thousand as of 1995, have labored to reforest and rejuvenate the land, planting over seventeen million trees.

This Green Belt Movement, now praised internationally, began on World Environment Day in June 1977 with a small ceremony involving a few women who planted seven trees in Maathai's own back yard. As a leader of an umbrella organization of women's groups, the National Council of Women of Kenya (NCWK), Maathai had been visiting with poor rural women, over 70 percent of whom are farmers. Casting about for a cost-free enterprise to help empower these women, give them hope, and provide short-term success as well as such long-term benefits as employment, Maathai came up with the idea of tree planting. Members of the NCWK established the first tree nursery behind the organization's office.

At first obtaining free seedlings from Department of Forestry nurseries, and collecting seeds from mature trees, the women nurtured the seedlings and then planted them on the land they farmed.

Over time, Maathai's army has transformed dusty, barren areas to green oases producing fuelwood, food, shade, medicinal substances, and oxygen, as well as bringing about water retention and soil enrichment. Under her tutelage, tree nurseries in Kenya—all locally managed—now number over three thousand. Children are able to eat their fill of fresh fruit right off the trees.

Maathai's work shows how traditional survival techniques such as intercropping and agroforestry can be inexpensively resurrected through women's practical skills. She says the Green Belt Movement "is committed to increasing public awareness of the relationship between environmental degradation and such issues as poverty, unemployment, malnutrition, and the mismanagement of natural resources, and the impact of these problems on the political and economic situation throughout Africa."[10] In 1987 the Green Belt Movement received the Global 500 Award of the United Nations Environment Programme.

Maathai's hometown, Nyeri, lies in the beautiful highlands of Kenya in a valley not far from Mount Kenya, the highest peak in the country. She says that before she left home in 1960 to attend college in the United States, her region, with its wild figs and other trees and its spring water, sustained the community. She saw no slums or starving children. When Maathai returned six years later, all the trees had been cleared for plantations, the local spring and stream had disappeared, and the land looked exhausted. Gullies and soil erosion, unknown before, scarred the earth. She saw hunger on the faces of the people. Nutritious food crops they had grown earlier had been sacrificed for cash export crops. And although the people then had a small amount of money, food was often not available in the markets and fuelwood to cook it was scarce.

Maathai—the first woman in East or Central Africa to earn a doctorate, the first woman to become an associate professor at the University of

Nairobi, and the first woman to head a department at the university—had received a full scholarship in 1960 to attend Mount Saint Scholastica College in Atchison, Kansas. She earned a master of science degree (again the first woman in East or Central Africa to do so) in biological science from the University of Pittsburgh in 1965, and then returned to Nairobi to obtain her doctorate. All this in itself was enough to rankle an entrenched patriarchal culture.

She launched the tree-planting project while teaching veterinary anatomy at the University of Nairobi and doing research on a cattle fever caused by a parasite. Her field work made her see that environmental degradation also posed a threat to the cattle industry.

After chairing the Department of Anatomy for seven years, Professor Maathai gave up her academic position in 1982 to run for Parliament, but a technicality disqualified her from being a candidate. She decided to concentrate on the Green Belt Movement, which by then had a small staff and had begun operating out of an office provided by the government.

As she describes it, "The Green Belt Movement harnesses local expertise and resources and encourages communities to stand on their own feet. We deliberately discourage direct participation by high-powered technicians and managers from outside. We want to create confidence in local people who are often overwhelmed by experts and come to think that they are incapable and backward."[11]

Before any tree plantings, Maathai and her coworkers engage the women in each community in a dialogue about desertification and about local issues and needs, eliciting their suggestions and learning from each other. Participants recite the Green Belt Movement pledge:

> Being aware that Kenya is threatened by the expansion of desert-like conditions, that desertification comes as a result of misuse of land by indiscriminate cutting down of trees, bush-clearing and consequent soil-erosion by the elements, and that these actions result in drought, malnutrition, famine and death, WE RESOLVE to save our land by averting this same desertification by tree planting, wherever possible.[12]

The women plant only such indigenous species as fig, banana, citrus, papaw, papaya, avocado, mango, nandi flame, acacia, thorn, and cedar—

never exotic species like pine or eucalyptus. Maathai teaches them always to plant more trees than they will need to cut for fuelwood. Given Kenya's mild climate, trees mature in about three to five years. For each tree that survives three months (80 percent do), the women receive a token payment of four cents, more money than most of them have ever earned. As their trees mature, they earn more by selling fruit and fuelwood.

At first funded by a Kenyan subsidiary of Mobil Oil and a few international foundations, the Green Belt Movement now receives almost all its financial support from outside Kenya in the form of small checks from women all over the world. Groups in Scandinavia have been especially generous. The United Nations Development Fund for Women provided a big boost in 1981 with a $100,000 grant. The movement's annual budget has grown to $500,000.

From the beginning Maathai took her tree-planting project to the schools. Children went to nurseries to fetch the seedlings, dug holes on school grounds, planted trees, and cared for them while finishing their studies. An average of five hundred students each in three thousand schools have participated.

The women's tree nurseries also give free seedlings to individuals and groups wishing to establish green belts. In addition to farms, homes, and schools, these gifts have transformed the grounds of hospitals, factories, and churches.

For over a decade the Kenyan government supported Maathai's Green Belt Movement—never financially, but with office space. In 1989, however, Maathai, an independent woman unafraid to speak her mind, angered government ministers, and especially the new authoritarian president, Daniel arap Moi, when she publicly opposed plans to construct a sixty-story office tower, hotels, theaters, a conference center, and a shopping mall in the middle of Nairobi's recreational Uhuru Park. (The president's plan also included a four-story statue of himself.) The ruling party admitted its motive to be mainly prestige—the office tower would be the tallest building in all of Africa.

Maathai filed suit against the ruling party on behalf of the Green Belt Movement—which had become a group of experienced activists—to prevent construction until the government produced an environmental impact report. Uhuru Park, one of the last remaining green areas in Nairobi and of historical significance, provided a major shady haven for downtown workers and urban residents, especially low-income persons living in crowded housing or in slums.

The country would have to borrow from foreign investors to pay for the $200 million complex, doubling Kenya's existing international debt. Worse, the proposed site lay over a potential earthquake area and an underground river. Although no politicians spoke out against the project, the Kenya Public Law Institute and the architectural society joined Maathai in opposing the plan.

However, a judge dismissed the suit, members of Parliament condemned Maathai (calling her action "ugly and ominous"), and the government denied her application for a permit to stage a public demonstration against the project. Furthermore, the government evicted her and the Green Belters from their ten-year-old office, giving them twenty-four hours to vacate. President Moi also threatened to deregister the group, making it illegal for it to function.

All this drew supportive attention for Maathai in the international media. Foreign investors backed off. A scaled-down plan for the tower, announced by the government early in 1991, never materialized.

Although the Green Belt Movement had won a significant victory, elevating the environment to a national political issue, it came at considerable cost. Without office space, Maathai and her staff members, along with ten years' worth of records, had to move into her house, which she then had to enlarge to accommodate the operation. During the renovation, filing cabinets stood in the driveway.

Meanwhile, Maathai's supporters abroad put pressure on the Kenyan government to ease up on oppression. As a result of their efforts, the World Bank and other donors delayed financial assistance to the country until political changes occurred that would allow an opposition party to form.

By this time, Maathai had concluded that restoring the environment would require more than planting trees. Basic human rights issues, a democratic willingness to allow dissent, and freedom from authoritarian rule were needed. So she allied herself with a new opposition party—the Forum for the Restoration of Democracy (FORD).

Learning of President Moi's plan to forestall an election and subvert a democracy movement by turning the government over to the military— after which opposition leaders were to be murdered—Maathai and eight other FORD leaders convened a press conference to expose the plot, which was to begin the following day. Police broke up the conference, but the crowd physically protected Maathai from arrest.

Later, however, 150 police stormed her house, which had been surrounded by her supporters. After a three-day confrontation, the officers smashed through windows and doors, arresting Maathai on charges of "spreading malicious rumors."

Already suffering from arthritis and heart problems, she spent the night in a damp, cold concrete cell without benefit of blanket or mattress. Maathai had to be carried on a stretcher to her hearing the following day. Although the police then released her, she needed to spend several days in the hospital recovering. Later, the government dropped charges. From then on, although not a fugitive, Maathai said she felt unsafe; she stopped going out alone or at night.

Not long after, in March 1992, a small group of older rural women relatives of political prisoners appealed to Maathai for help. They had begun a hunger strike in Uhuru Park to gain release of their sons. Maathai could not refuse, since these women exemplified those she strove to empower. Soon hundreds of sympathizers joined the protest, and leaders of the new opposition party took advantage of the occasion to deliver speeches. The government dispatched truckloads of police to the scene. They attacked the crowd with tear gas and then with clubs. Maathai, trampled and beaten unconscious, awoke in the intensive-care unit. She immediately arranged a press conference from her bed, demanding release of the political prisoners.

News of this event and photos of the battered Maathai flashed around the world. Activist groups everywhere publicly lauded Maathai's courage. Protest mail flooded President Moi's office. He released the political prisoners. Maathai gained international recognition, an important step in opening the way for women in Kenya to participate in leading the new opposition party.

Although her passport had been impounded by the government, Mikhail Gorbachev intervened on Maathai's behalf, enabling her to attend the 1995 meeting of his International Green Cross Society in Tokyo. Since then, she has experienced less harassment.

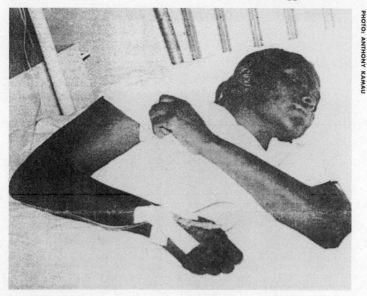

PHOTO: ANTHONY KAMAU

WANGARI MAATHAI IN HER BED AT THE NAIROBI HOSPITAL.

While teaching at the University of Nairobi in the 1970s, Maathai had married. She helped her husband be elected to Parliament in 1974. They had three children, two boys and a girl. But her husband had only a bachelor's degree. African culture expects women to be submissive and dependent, in no way "better" than their husbands. Unable to accommodate his wife's advanced academic achievement, Maathai's husband filed for divorce, falsely accusing her of adultery. She fought the action at the court hearing, suggesting that the judge must be either corrupt or incompetent to rule in favor of a divorce based only on rumors he had heard. She landed in jail for contempt of court. Her lawyers secured her release after she had served three days of a six-month sentence. Conditions of her release included an apology to the judge.

Some years later she remarked that not having a husband had been a blessing, that she could not have accomplished all she had with a husband—at least not the one she had married.

Now considered an eco-hero internationally, Maathai has received (besides two honorary doctorates) a number of awards. These include the Right Livelihood Award (1984), the first *Condé Nast Traveller* Award

(1990), the Goldman Environmental Award (1991), the fifth City of Edinburgh Medal (1993), and the Golden Ark Award from HRH Prince Bernhard of the Netherlands (1994).

Maathai has been an active board member of several organizations, among them the Environmental Liaison Centre International, the WorldWIDE Network of women in environmental work, and the National Council of Women of Kenya. She also serves on the Sasakawa Environment Prize selection committee. In 1991 she actively participated in the World Women's Congress for a Healthy Planet, held in Miami to establish a proposed agenda for the United Nations Conference on Environment and Development (UNCED)—also known as the Earth Summit—that took place in Rio de Janeiro in 1992. As originally proposed, the UN conference declaration on the environment had almost no references to women. As part of a caucus of women's organizations, Maathai pushed for change; this led to the addition of 120 provisions affecting women. In 1995 she participated in the Fourth UN World Conference on Women in Beijing. Now in demand as a speaker at international conferences, she has traveled and been interviewed extensively. In 1988 she wrote a book about the Green Belt Movement, and her articles have appeared in scientific journals and popular magazines.

Through Maathai's work with nongovernmental organizations (NGOs) elsewhere, the Green Belt Movement has spread to more than thirty countries. It has earned the reputation of being one of the world's most successful projects in bringing about environmental restoration combined with community economic development.

Summing up, Maathai says, "My greatest satisfaction is to look back and see how far we have come. Something so simple, but meaning so much, something nobody can take away from the people, something that is changing the face of the landscape. . . . I never knew when I was working in my backyard that what I was playing around with would one day become a whole movement. One person can make a difference."[13]

COLLEEN McCRORY (1953–)
Crusades for Canadian Forests

Mention rain forests and the tropics come to mind. Yet a unique two-thousand-mile temperate rain forest in North America stretches along the Pacific Northwest coast from Alaska to northern California. With 180 inches of rain a year, it produces ten times more vegetation per acre than a tropical rain forest. In addition to providing critical wildlife habitat for thousands of species, this Pacific Northwest rain forest, through transpiration, creates almost one-third of Earth's total rainfall.[14] The forest functions like a huge sponge, taking up and retaining moisture. Just one old tree can contain thousands of gallons of water in its trunk, branches, and roots. Old-growth forests comprise many components: large living trees, large dead trees (both standing and fallen), and numerous layers of vegetation. A vast community of insects, birds, plants, and mammals depends on various parts of each old tree. Even after a thousand-year-old tree dies, it will nurture forest life for another four centuries. Humans may plant trees, but they cannot re-create a natural forest. British Columbia, the most biologically diverse region of Canada, shelters 70 percent of the country's bird species and 74 percent of its mammals. Its ancient stands of trees are among the earth's last few remaining primary forests. But British Columbia, with logging its primary industry in the twentieth century, has already cut 60 percent of its timber (in the United States, the figure is 90 percent).[15] During the past thirty years alone, the amount of forest cut in the province has tripled, with the rate in the late 1990s being one acre per minute.

When local loggers hurled a rock through her Valhalla Trading Post window, threatened her children at school, and organized a boycott of her store, it would have been easy for Colleen McCrory to give up her cause and move away. She almost did. The angry loggers, not understanding the real reasons why their jobs were at risk, targeted environmentalists as the enemy.

McCrory had lived in New Denver, British Columbia (population less than a thousand), all her life, knew everyone, and had considered them all friends. She and her eight siblings grew up playing in the foothills of the forested Valhalla Mountain Range, which lies between British Columbia's Pacific Coast Mountains and the Rockies to the east. Encompassing typical British Columbia landscape, the Valhalla Range is part of the Pacific Northwest temperate rain forest.

In 1975, concerned about the destruction of Canada's forested wilderness areas, and driven by a desire to do something about it, McCrory organized a group—the Valhalla Wilderness Society—to inform the public about the value of the Valhalla Range and lobby for its protection. A high-school dropout, unfamiliar with the technical jargon of the timber industry, McCrory educated herself as she went about lobbying for the creation of a provincial park. To gain broader media coverage, she traveled to Vancouver, a fourteen-hour bus ride from New Denver, where she knocked on the doors of radio and television stations, telling her story. Next she lobbied cabinet ministers in Victoria, British Columbia's capital, making the long trip every couple of weeks; sometimes she had to bring along her youngest child, then still a toddler.

McCrory, along with the directors of the Valhalla Wilderness Society, lobbied the provincial ministers for eight years. They created a traveling show of striking visuals, music, and narrative that toured the province; their television appearance on the Canadian Broadcasting Corporation generated a deluge of letters from around the country. There was a constant stream of well-researched news releases exposing the misleading statistics of the government and the logging company. McCrory attributes the success of this campaign to always following through, never taking time off, remembering the importance of attending meetings, answering phone calls and letters, and appearing on radio and television. In the end, though, the massive support for preserving the Valhallas rested on the fact that the timber from the entire mountain range made up only 3.4 percent of the logging company's cut. Finally, in 1983, the government created the Valhalla Provincial Park, preserving three-quarters of the range facing Slocan Lake. McCrory believes it is comparable in beauty to the well-known Banff National Park.

McCrory had left high school after the eleventh grade to marry. Three children later, and after becoming involved in saving the Valhalla Range from the chainsaw, she left the marriage, thereafter devoting all the time she could to helping save Canada's ancient forests from overexploitation.

Not knowing how she would support herself, McCrory and her children moved into a lakeshore house in New Denver heated only by a woodstove. Along with her ongoing lobbying, she opened a small store in New Denver, the Valhalla Trading Post, where she sold a few books, camping supplies, jewelry, and clothes.

Even before the Valhalla Provincial Park became a reality, McCrory's reputation as a conservation activist had spread across Canada. She had acquired lobbying, media, public-speaking, and fund-raising skills and had established professional relationships with decision makers in government.

McCrory feared that at the rate the timber industry was going the ancient forests of British Columbia faced destruction within a couple of decades. Even though she and her colleagues in the Valhalla Wilderness Society were exhausted, and the society was in debt following the eight-year struggle to win the provincial park, McCrory realized that the lessons learned in that effort could be applied to other environmental causes. So she began helping in the battle to establish a national park in the Queen Charlotte Islands.

These islands lie about sixty miles off the northern Pacific coast of British Columbia, over a thousand miles away from McCrory's home. She had never been there, but had obtained photographs and detailed information about the islands' unique ecological features, including plants and animals, thousand-year-old trees, and eleven species of whales. Environmentalists sought to have the southernmost section of the Queen Charlotte Islands (called South Moresby) designated a national park reserve. Achieving this goal required lobbying in Ottawa, Canada's capital, some two thousand miles to the east.

McCrory had not been there, either. But her chance came when she traveled to the Atlantic Coast to accept the Governor General's Conservation Award. During a scheduled stopover in Ottawa on the return flight, she left the plane to seek an audience with the federal environmental minister. By the end of the day she still had not spoken with him. Having missed her flight home, she refused to leave Ottawa until she talked with the minister about South Moresby. Having no place to stay overnight, she went to the home of the member of Parliament from her district, in whose election campaign she had worked. She persuaded one of his children (he was not at home) to let her sleep on a sofa.

Late the following evening she met with the environmental minister and presented her case for designating South Moresby a national reserve. In describing the area, she characterized it as Canada's Galapagos. The minister wanted to know how much it would cost. McCrory, not prepared to answer that question, made a guess of twenty million dollars, an estimate she found out later was close to being accurate.

The minister then agreed that it could be done. During the three years of negotiations between the provincial and federal governments that followed, a national election brought in a new administration. Anticipating such a change, McCrory had obtained a commitment from the political party likely to win, confirming its support for a South Moresby National Park Reserve. In 1987, the federal and provincial governments signed the agreement officially establishing the park.

Meanwhile, McCrory had difficulty paying her living expenses, sometimes facing disconnection of telephone and electric service for non-payment, a common situation among grassroots activists, who often lead hardscrabble lives.

The creation of South Moresby National Park riled the logging companies. The editor of a publication called the *Red Neck News* in Sandspit, a pro-logging community in the Queen Charlotte Islands, launched an attack on McCrory and the Valhalla Wilderness Society. Articles in his paper falsely accused her of terrorism and vandalism, even mentioning arson. The writer of the article publicly admitted being paid by a logging company. The vitriolic abuse incited the Valhalla logging interests to organize a boycott of McCrory's store. Men and women she had associated with all her life turned against her. "I could feel the scorn and anger in my community. At social functions angry loggers accused me of taking away their jobs," she recalls. The boycott dragged on for two years, forcing McCrory to close down and leaving her with a fifty-thousand-dollar debt in a hostile community.

Though at this point she felt like a failure and it crossed her mind to leave town, McCrory's sense of obligation to defend the environment stiffened her resolve. A gregarious person from a large family who had always enjoyed friendly associations with many people, McCrory says that "the social ostracism caused me more pain than my financial plight." Then, a ten-thousand-dollar donation from someone she did not even know helped restore her faith. In addition, a nonprofit environmental organization, the World Wildlife Fund, raised money for her.

With unemployment high throughout British Columbia, McCrory sympathized with the loggers' fears about their livelihood. But the destabilization of the timber industry can be traced, not to environmental measures promoting sustainable logging, but to such factors as diminishing the resource through overcutting, internal corporate practices and mismanagement, market changes, and evolutionary technological advances like automation. Environmentalists' efforts to preserve the forests actually help prevent economic collapse.[16]

An independent study reported that only 2 percent of the twenty-nine thousand forest-industry jobs lost between 1981 and 1991 disappeared because of environmental-protection efforts.[17] "Many forest products processing jobs that could stay in British Columbia," McCrory claims, "end up overseas, mainly in Asia." She advocates diversifying the province's economy instead of continuing to rely on timber. She believes a healthy natural environment stimulates job creation. Over the past few years wilderness tourism in British Columbia has increased by almost 40 percent and has become the fastest-growing local industry. According to a government study, logging costs in the Valhalla region outstrip profits. The study suggested that, by contrast, protecting the region would mean $16 million in new capital investment opportunities, annual revenue from tourism amounting to $3.4 million, and 229 permanent jobs.[18]

To avoid taking a piecemeal approach to preserving Canada's forests, McCrory and the Valhalla Wilderness Society brought together a coalition of four hundred groups, representing a million Canadians. It developed a map of the country, with the areas proposed for protection colored green. Canada's Future Forest Alliance, as the coalition is called, proposes that the current 5.24 percent of the country's land base designated as parks be increased to 13 percent, the figure suggested by the UN for each nation. Under this plan the percentage of Canada's forests excluded from logging would be around 5 percent instead of the current 2 percent. Creating the coalition's plan required five years of research, which was led by McCrory and the Valhalla Wilderness Society. The plan proposes 122 new parks, which would include the last large wilderness areas in the world.

The boreal forests of the far north, ringing the Arctic Circle, have been targeted by the timber industry for logging. While on a speaking trip in northern Alberta, McCrory learned of plans to construct forty new paper-, pulp-, and sawmills throughout Canada's boreal forest areas.

Without environmental impact studies, public hearings, or consideration of the communities affected, some of the largest pulp mills on the globe—with thousands of miles of accompanying logging roads—are being built.

Boreal forests, which make up 25 percent of the earth's remaining forests, consist of pine, spruce, fir, aspen, birch, poplar, and larch trees. Because the soil around their roots thaws for such a brief time each year, the trees grow slowly. If they come back at all after cutting, it can take a century or more for a boreal forest to regenerate. Pulling carbon dioxide from the air, storing it in organic material, and generating oxygen, these trees perform a critical function in helping prevent global warming. Experts believe these northern forests store more carbon than all the world's tropical rain forests put together. But exposure of the soil to sunlight and air after logging causes the release of carbon dioxide into the atmosphere, exacerbating the greenhouse effect.[19] Recent summer temperatures in northern Siberia have been the warmest in a thousand years, threatening its boreal forests by forcing them to migrate farther north.

Eyeing these unique stands of trees, the Canadian logging industry plans to double its activities in the last years of this century. McCrory and Canada's Future Forest Alliance seek a moratorium on new "forest development" projects.

Meanwhile, McCrory continues to travel, speak, and educate citizens and policy makers on the importance of protecting forests. She jokes about sleeping on just about every couch in Canada during her travels. During visits to Brazil and Japan, she became involved in organizing the Taiga Rescue Network, an international campaign to preserve the world's boreal forests.

McCrory believes that the cozy relationship between governments and multinational corporations siphons off a great many tax dollars—money that could be invested in creative and profitable enterprises in local communities.

McCrory never quits. Still in the trenches after more than twenty years as an eco-warrior, and now a grandmother, she is leading a campaign in British Columbia to protect the Slocan Valley watersheds from clear-cutting. These watersheds, considered one of the world's rarest ecosystems, ensure clean drinking water for residents of the region.[20] Disregarding opposition to logging the area by 97 percent of the local citizens, the British Columbia government has awarded pork-barrel contracts to a timber company to clear-cut the valley watersheds. Citizens, believing that the government would abide by an earlier pledge to respect a forest management plan worked out by the local communities, engaged and paid the Silva Foundation to prepare a land-use plan based on the needs of the ecosystem. The government then ignored the plan.

McCrory, working with the Rainforest Action Network, persuaded representatives of twenty-three international environmental groups to sign a declaration at the Rainforest Summit in London in September 1996. The statement threatens to organize a worldwide boycott of Slocan Forest Products—the giant timber company given the contracts—if the destruction proceeds. McCrory, who received one of the Goldman Environmental Prizes in 1991, says she will fight logging the Slocan Valley area even if it should lead to her arrest.

> After twenty years of effort protecting the watersheds and old-growth forests, we're not about to sit back and let [the current provincial government] do what previous . . . governments wouldn't dare. When push comes to shove, the British Columbia government shows its true colors. It would sooner bulldoze democracy than stand in the way of its beloved timber interests. The only people who want this logging to happen [are in] the logging company itself. It's making democracy a farce when all that matters is corporate profits.[21]

How the Slocan Valley confrontation will play out remains uncertain as of early 1998, with citizen protests and court battles ongoing.

Almost half of the prime timber cut down in British Columbia is exported to the United States for use in throw-away telephone books,[22] newsprint, magazine stock, and other kinds of paper, as well as for such items as cedar shakes and decking. McCrory pleads with American consumers to buy and use only 100 percent recycled, unbleached paper products and to become involved in grassroots efforts to save the remaining fragments of the earth's primary forests.[23] Meanwhile, she and the Rainforest Action Network campaign to redirect the use of wood products in the United States away from British Columbia's temperate rain forest to ecologically sound alternatives.

McCrory's example demonstrates what persistence—even against overwhelming opposition—can achieve.

JUDI BARI (1949–1997)
Redwood Warrior

Three hours after a bomb hidden under the driver's seat of her car exploded, the Oakland police arrested Judi Bari in surgery. When she regained consciousness in the intensive-care unit, FBI agents charged her with transporting illegal explosives. Though she had a leg in traction, tubes trailing from her body, a pelvis shattered in twenty-seven places, and a paralyzed lower body, the police declared her a "flight risk" and raised her bail from the usual $12,000 to $100,000. At first, while under protective custody, she was forbidden to have visitors other than her family or to have physical contact with anyone. Even her daughters, then aged four and nine, could not hug their mother.

Bari—a commercial carpenter by training, and a musician, labor unionist, environmentalist, writer, radio host, and Earth First! activist as well—had emerged during almost a decade of involvement in Earth First! as a skilled organizer, strategist, and thinker. The group had undertaken to help save the remaining ancient redwood forest ecosystem of northern California, especially an area called Headwaters Forest, which contained the earth's last large stand of unprotected ancient redwood and was home for 160 wildlife species. The trees (some of them two millennia old, almost the age of the Great Wall of China) are part of the Pacific Northwest temperate rain forest. Some measure more than fifteen feet in diameter, and just one tree can be worth more than $100,000. Ferns on the surrounding forest floor grow taller than people.

The redwood forests provide the nesting habitat required by the endangered spotted owl and by the marbled murrelet (threatened according to the federal list, but endangered in California), a rarely seen seabird that nests in the forests' topmost canopies. The National Biological Service reports that this bird's population has suffered a marked decline during the past few decades.[24] More than 97 percent of California's old-growth redwoods have already been cut. Only about sixty thousand of the original two million acres remain.

Since its founding in 1869, the family-owned Pacific Lumber Company (owners of the Headwaters Forest) had practiced sustainable logging in its redwoods holdings. Then, following a hostile takeover (financed by junk bonds) by Maxxam, a corporation based in Houston, the company tripled the logging rate, leveling forty thousand acres of ancient redwoods.

Earth First!, a nonhierarchical citizens' environmental movement, employs a front-line, direct-action approach. Tactics range from rallies and demonstrations to blockades, pranks, tree sittings, and (like the Chipko protesters in India) putting their bodies in the way of moving bulldozers. The movement also engages in civil disobedience intended to lead to arrests.[25]

Although some Earth First!ers, especially the founders in the early 1980s, practiced tree spiking (potentially fatal to sawyers) and "monkey-wrenching" (such as pouring sand into bulldozers' gas tanks), Bari opposed both these actions and led the effort to transform the organization's tactics. In April 1990, a month before her car exploded, Bari and several other Earth First! leaders issued a memorandum to all associated groups and individuals urging them to renounce tree spiking, at least in northern California and Oregon.[26] "The reason," she wrote, "is that Earth First! has been so successful in working and strategizing with timber workers that the alienation caused by tree-spiking, not to mention the danger, be it real or imagined, was harming our efforts to save this planet. . . . The loggers and millworkers are our neighbors, and they should be our allies, not our adversaries. Their livelihood is being destroyed along with the forests. The real conflict is not between us and the timber workers, it is between the timber corporations and our entire community."[27] Having worked as a laborer herself, she could identify with millworkers.

Bari grew up in Maryland and attended the University of Maryland, where she joined in protesting the Vietnam War. Only five feet tall, she earned a karate black belt and could hoist a seventy-pound mailbag. A blue-collar worker after dropping out of college, she organized unions and led strikes. Described as funny, impish, sharp-tongued, smart, scrappy, and strong-willed, Bari, having gained a reputation as a workers' advocate after moving to northern California, had begun helping mill-workers force the timber companies to correct hazardous working conditions. For example, workers at a Georgia-Pacific plant had been exposed to PCBs during a large spill of contaminated oil. She taught them how to file complaints with the Occupational Safety and Health Administration (OSHA) and how to keep records of hazards.[28] She helped millworkers organize a new union when the existing one failed to protect their health and safety. Bari thus gained allies inside companies whose managements were bent on sacrificing the redwoods for dollars. She also convened regular meetings between timber industry employees and environmentalists. So why did a bomb explode in her car?

Judi Bari believed she was a target because of her "rabble-rousing." She had become a highly visible and outspoken leader of Earth First!'s increasingly successful redwoods protection campaign. Through more than a dozen lawsuits filed by an ally—the Environmental Protection Information Center (EPIC) in Garberville, California—and through tree sittings, blockades, rallies, demonstrations, and arrests for civil disobedience, Earth First! aroused national awareness of the redwoods issue.

In the months before the bombing, Bari, on a leave of absence from her carpentry job, had been organizing for Redwood Summer, a massive drive intended to take place from June to August in 1990. Modeling it after the Mississippi Summer civil rights campaign of 1964, she anticipated that several thousand young people from all over the country would congregate in northern California to take part in Earth First!'s redwoods crusade. The participants would be required to undergo two days of non-violence training. When the bomb exploded, the fiddle-playing Bari and another Earth First! organizer and musician were on their way to Berkeley to make final plans for Redwood Summer with the pacifist group Seeds of Peace, which provides food for large nonviolent gatherings.

During a three-week period not long before the bombing, Bari had received several death threats. For example, a blown-up newspaper photograph of her, with rifle-sight crosshairs drawn over her face, was tacked to the door of the Mendocino Environmental Center in Ukiah, where she spent a lot of time.[29] When she got in touch with the Ukiah police, they replied that they lacked the personnel to investigate such threats, adding that if she turned up dead they would look into the matter.

A day after having been blockaded by activists, a timber worker followed Bari in a log-hauling truck and rammed her car, sending it sailing off the road. Fortunately, her children received only minor injuries, but the crash demolished her car. In addition to telephone threats, she received a typewritten message that read: "Get out and go back to where you came from. We know everything. You WILL NOT GET A SECOND WARNING."

Others also received threats. Bari had succeeded in persuading some of the members of the Mendocino County Board of Supervisors to consider using its power of eminent domain to take over ancient-forest acreage owned by the timber company. One of the supervisors, after a one-on-one follow-up meeting with Bari, heard this message on his home answering machine: "You will be killed."

In an article entitled "Misery Loves Company," Bari detailed examples of violence against environmental activists and whistle-blowers: murders, homes burned, dogs killed, drive-by shootings. She said, "Threats and intimidation are becoming common tactics against all kinds of environmentalists. We shouldn't have to take our lives in our hands to stand up against poisoning a creek or over-cutting the forest. But the corporations and police are all too willing to use excessive force against us."[30] She believed that the FBI and local police "encouraged vigilantes by sending a clear message that crimes against Earth First!ers will not be prosecuted, including the bombing of me."[31]

Two months after her car was bombed, the FBI and the Oakland police, after raiding Bari's home twice, found that they had no evidence to substantiate their accusations. They dropped the case against her. One of the FBI agents who had arrived at the bomb scene within minutes—special agent Frank Doyle—claimed in a search-warrant statement that the bomb had been located on the floor of the car's back seat in plain sight. Photos by both the Oakland police and the FBI's bomb expert in Washington, D.C., however, showed the three-foot by five-foot hole in the car's floorboard to be directly under the driver's seat.[32] Moreover, the FBI's bomb expert later testified that the bomb had two triggering mechanisms: a twelve-hour timer that first had to expire, followed by a ball-bearing motion device that would set off the bomb when Bari drove the car.

Bari never knew who planted the bomb. But, she said, "I know who did the arrests and who did the cover-up of the bombing and who did the false vilification of me . . . and Earth First!, and that was the FBI."[33] In the 1980s the FBI, with the complicity of the local police and the media, had been working with anti-environmentalist individuals and groups to brand Earth First! as terrorist. The tactics paralleled those employed by the FBI's counterintelligence program (COINTELPRO) against civil rights activists—a program that a Senate committee found unconstitutional in 1975. Bari reported that Richard Held, the FBI agent in charge of her bombing investigation, "became notorious in the 1970s for his active role in COINTELPRO, an outrageous and illegal FBI program to disrupt, misdirect, isolate, create a climate of fear, neutralize and destroy any group that challenged the powers that be."[34] She added, "I cannot describe the cold terror of waking up in the hospital, crippled for life, and finding out that Richard Held was accusing me of blowing myself up with my own bomb. . . . Not just the FBI, but now private corporations as well, are using the same counterintelligence tactics that were declared illegal in the 1970's to try to discredit and destroy Earth First! in the 1990's."[35]

Bari believed rogue FBI operatives or timber company anti-environmentalists infiltrated Earth First! to win trust and establish friendships. They then issued fake press releases on stationery displaying the Earth First! logo, calling for violence against timber workers. These were widely distributed to millworkers and to the local press. Though acknowledging in an internal company memo that the press releases were phony, the Maxxam/Pacific Lumber Company—through its public relations firm, Hill and Knowlton—released one of them to out-of-town newspapers.

The infiltrators also engaged in dirty tricks intended to depict Bari as a terrorist. She told about a man who, making himself out to be an Earth First!er by participating in rallies, talked her into letting him—for a lark—take a photo of her holding a modified Uzi submachine gun he kept in his car trunk. He showed her how to hold it and posed her for the photo to display her Earth First! T-shirt. Bari learned later that he had sent this joke photo to the police, anonymously, with a letter saying he had joined Earth First! to report illegal activities, mentioning Bari as a leader and main force of Earth First!, falsely accusing her of planning vandalism against the property of a member of Congress, and alleging that Earth First! was engaging in automatic-weapons training. The police, after Bari's car was bombed eighteen months later, released the photo to the media. It appeared in mainstream newspapers as "proof" that Bari was a terrorist. However, she was able to establish, through document analysis equipment, that the man who took the photo, using the same manual typewriter as the one used for the anonymous letter to the police, had composed one of the death threats she received. Bari claimed the FBI spent three million dollars on infiltrating and disrupting Earth First!.[36]

Two weeks before she was almost killed, the FBI had conducted a bomb response training class for law enforcement personnel at the College of the Redwoods in Eureka, California. In a clear-cut timber area belonging to the Louisiana-Pacific Lumber Company, those attending assembled and detonated the same kind of complex pipe bomb as the one used in Bari's car. She later inquired how many of the respondents at her bomb scene had taken the training class. Neither the college nor the FBI would give her a copy of the roster. She appealed to the courts without success to force release of the document.

In spite of the threats and violence, Earth First! activists did not back away from their campaign to save the redwoods. While Bari recovered in the hospital, Redwood Summer went on. Young people, mostly women, rose to leadership roles in her place. "Three thousand people from all over the country came to Redwood Summer [in 1990] and chained themselves to logging equipment, hugged trees, blocked logging roads and marched through timber towns,"[37] making the issue of redwood slaughter one of national and international concern.

Bari herself, though disabled and in constant pain, continued her activist leadership in the years following the bombing. During his announcement in September 1995 that the company had obtained permission to do "salvage logging" in Headwaters Forest, the president of Maxxam/Pacific Lumber stated that only fifty or sixty persons cared about the redwoods. Following a rally with two thousand participants on September 15, Bari and a smaller group of two hundred activists stood at the gates of Maxxam/Pacific headquarters with written evidence proving him wrong. During the preceding year, the Bay Area Headwaters Coalition had collected twelve thousand signatures on petitions opposing the logging of Headwaters Forest.

As one of three persons designated to present the petition to the company's president, Bari (bullhorn in hand) occupied a position in the front row of the group. When a security guard refused to allow the three presenters to enter, Bari aimed her bullhorn at the second floor executive offices and chided the president for being too busy hiding under his desk to accept the petition. At about that time, company provocateurs tried to bait the group by scuffling with the guard. "The crowd stood firm and steady, chanting. 'No violence! No violence!' And each time the company thugs got preoccupied with the provocateurs, [we] took another step forward, closing the gap between us and the door. It was one of the most powerful moments I have witnessed, as the nonviolent crowd created a situation in which the company's violence had no context."[38] Bari told the president through the bullhorn that all he had to do to get rid of them was to allow three representatives to respectfully deliver their petition. " 'Otherwise, we're not leaving.' And amazingly, John Campbell [company president], the most powerful man in Humboldt County, backed down."[39] Later the same day, when a contingent of activists collected at the forest and in deliberate civil disobedience crossed over onto timber company property, Bari headed the line of 264 protesters waiting to be arrested. The protest resulted in another logging postponement.

The smear campaign against Earth First! did not cease. In April 1996, ABC's Peter Jennings on *World News Tonight*—relying only on a timber industry informant who wrote a book about how he infiltrated Earth First!—aired a report trying to link the group with Theodore Kaczynski, alleged to be a terrorist known as the Unabomber.[40]

So what will happen to the redwood forests, especially the Headwaters Forest, that Judi Bari, Earth First!, and other public interest groups worked to save? Late in 1996 California and the federal government reached an agreement whereby Maxxam/Pacific Lumber will halt logging on seventy-five hundred acres in the heart of the Headwaters Forest, ceding it to the federal government and California for a preserve in exchange for $430 million in assets. Environmental activists say this acreage is not enough and continue to press for preservation of the entire sixty-thousand-acre watershed.

On the legal front, the U.S. Supreme Court in February 1997 ruled unanimously to let stand a lower court ruling barring logging in Headwaters Forest, even though the logging company owns the land.[41] In 1993 the Environmental Protection Information Center (EPIC) of northern California had filed suit in federal court against Maxxam/Pacific Lumber for destroying marbled murrelet habitat in violation of the Endangered Species Act. In February 1995 the court ruled in favor of EPIC and issued a permanent injunction barring logging. All of the lumber company's appeals failed.

In the meantime, a coalition of environmental groups continued to carry out creative actions to protect redwoods from the chain saws. In February 1997 environmental leaders teamed with entertainment celebrities to announce a nationwide campaign, inaugurated by a full-page advertisement in the *New York Times*, urging consumers not to buy old-growth redwood products.

In 1991, Bari had filed suit against the Oakland police and the FBI for false arrest, illegal search and seizure, civil rights violations, and denial of equal protection of the law. Her attorney claims the two law enforcement groups falsified evidence, covered up their wrongdoing, and lied to the media. He says some of the lies involved hanky-panky in the FBI's now infamous crime lab.[42] The bureau claims immunity from such a suit. But the courts rejected its four attempts to have the case dismissed, and, as of early 1997, the matter appeared to be going to trial. Through the Freedom of Information Act, and during three and a half years of the "discovery" phase of the case, Bari gained access to seven thousand pages of FBI and Oakland police files, plus six thousand pages of sworn testimony from individuals in both organizations and from other witnesses. However, critical sections of the FBI documents had been blacked out, she said, and the bureau withheld other documents to which she was legally entitled.

FBI agents had repeatedly claimed that they had never heard of Judi Bari before the bombing. "Yet the Oakland Police have said in their sworn testimony that the FBI appeared at the bombing scene and told the OPD that Darryl [the passenger in the car] and I were known terrorists, that we were Earth First!ers, and that we were the 'type of individuals who would be involved in transporting bombs.' " The bureau's own documents show, Bari said, that the "FBI was conducting undercover operations against Earth First! in Mendocino County, where I live, as early as 1988. . . . [The report] shows the FBI engaging in illegal spying, targeting political activists in phony terrorism probes." She emphasized that this case was "about the right of all activists to work for social changes without fearing repression by the government's secret police."[43]

Bari told a colleague at the Mendocino Environmental Center that if they wanted to stop her they bombed the wrong end. Doctors predicted she would never walk again. But somehow she made her numb leg and foot work well enough to move around, and she became the host of a monthly local radio show. She also began to write articles. In response to requests for copies of her pieces chronicling her experiences, she collected her writings in a book, *Timber Wars.*

In November 1996, Judi Bari disclosed that she had an aggressive, inoperable form of breast cancer that had already spread to her liver. Her doctors gave her less than six months to live. She died on March 2, 1997, at her remote "hippie shack" on the wooded slopes outside Willits, California. She had requested that in her obituary her occupation be designated as "revolutionary." One week after her death, over a thousand persons took part in a memorial celebration of her life. Her friends and family intend to pursue her lawsuit against the Oakland police and the FBI.

HARRIET BULLITT (1924–)
Invests in Earth's Future

Dividing her time between a cottage alongside a river in Washington's Icicle Canyon and a fifty-year-old wooden tugboat, Harriet Bullitt has chosen a lifestyle of voluntary simplicity. Eschewing ostentation, she feels that simplicity is luxury. Her brand of environmental activism has its own special character. Her family foundation gives away about four million dollars a year to a wide range of environmental causes, most often forest protection groups. Additional contributions flow from Harriet Bullitt's personal trust. She says she gives from self-interest, making it possible for others to do the actual work of protecting and restoring the planet for her own and everyone's grandchildren.

Hiker, kayaker, cross-country skier, fencer, flamenco dancer, and tugboat captain, Bullitt—at seventy years of age and newly married to a Russian emigré almost four decades her junior—embarked on a new career as innkeeper of a conference center and retreat across the Icicle River from her own property. She purchased a onetime Civilian Conservation Corps (CCC) camp—which lies in a pocket of the forested north central Cascade Mountains near Leavenworth, Washington—to prevent its being commercially developed.

It came on the market in 1992, soon after she experienced an inflow of capital from two sources—the sale in 1988 of a publishing company she had founded in the sixties, and the sale in 1991 of the family's 70 percent interest (amounting to about $80 million) in a regional media business, King Broadcasting. Most of the proceeds, along with a sizeable portion of her mother's estate, were funneled into the family foundation earmarked for environmental causes.

Harriet Bullitt's commitment to protecting the natural world began early. When she was a child, Icicle Canyon became her year-round vacation playground after her mother built a family retreat there in the 1930s—a rustic lodge they called Coppernotch. Bullitt rode ponies along the banks of Icicle Creek and hiked the surrounding slopes. She says the water, the forest, and the mountains are part of her life and always have been. As a grown-up she has escaped to the place whenever she needed to be away from public activities, make important decisions, or write articles for her magazine. Even when she traveled to the opposite side of the globe, she came back to Icicle Canyon mentally. "This has been my spiritual and material home—my one continual home—my entire life." Bullitt still grieves over the disappearance fifty years ago of the masses of salmon species that had thrived in the creek before encroaching agriculture, development, and dams on the nearby Columbia destroyed its natural flow.[44]

❧ Bullitt's higher education endured a series of interruptions. But in 1965, after her children were grown, she earned her bachelor's degree in zoology from the University of Washington. She had first registered there in the early forties as an engineering student. Forbidden to use the engineering library (because she would "distract the male students") and losing interest in her courses, Bullitt transferred to Bennington College in Vermont. She dropped out in 1945 to marry and raise two children. During her family-oriented years, she managed to work at three medical research jobs—in physiology, bacteriology, and protein chemistry—as a lab technician.

Living outside the Pacific Northwest for a while, and observing on her return runaway growth, worsening air and water pollution, clear-cutting, and Puget Sound's dwindling populations of sea creatures and other species sharpened Bullitt's environmental consciousness. Believing that if men and women better understood and appreciated the natural world they would treat it with more care, Bullitt in 1966 started a natural history publication called *Pacific Search* and established a publishing company. At first only a newsletter circulating to a few thousand subscribers, it evolved into a popular glossy periodical, *Pacific Northwest Magazine*, reaching seventy-five thousand subscribers each month. Long before it became fashionable to do so, and against the advice of her editor when the journal needed the revenue, Bullitt refused to accept tobacco advertising. The publication was once nominated for the coveted National Magazine Award for General Excellence. Bullitt sold her publishing venture in 1988.

Though they had an affluent life while growing up during the Great Depression, Harriet Bullitt, her brother, Stimson, and her sister, Patsy Bullitt Collins (who share her commitment to environmental causes) recall learning as children the basic values of thrift and resource conservation. Household rules dictated that nothing be wasted: no food left on plates, no leftovers uneaten, no clothes lost. Everything from household items to toys underwent repeated repairs and reuse. The siblings learned that having money did not make them better than others, and putting money back into the community was a given. Today the sisters and their brother share the Bullitt Foundation's decision making about the many grant requests they receive.

The family fortune grew from the logging business in the late 1800s and early 1900s. Successful investments in Seattle real estate added to the wealth, which mushroomed with the success of the King broadcasting empire pioneered by Bullitt's widowed mother, Dorothy Stimson Bullitt, in the 1940s.

Northwest timber company executives, angry about the Bullitts' giving money to nonprofit environmental groups focusing on forest preservation, view them as traitors and accuse them of acting out of guilt. Others consider the two sisters eccentric, wealthy meddlers not capable of identifying with workers who believe that environmental measures threaten their jobs.[45] The Bullitts remind loggers that when their one-armed grandfather was in the timber business, cutting down a single large tree required an entire day's work. There were no bulldozers and chain saws. He did not rape the land and quickly move on. The Bullitts today feel no guilt.

The Bullitt Foundation does not shrink from controversy. For example, between 1991 and 1996 it awarded $380,000 in grants to the Sierra Club Legal Defense Fund to support legal challenges to logging in spotted-owl habitat and to protect and restore salmon habitat.[46] In addition to awarding financial grants to protect natural resources, the foundation offers management- and technical-counseling sessions in an effort to balance economic growth and environmental values. The foundation avoids blanket opposition to every kind of development. It gives priority to identifying environmental problems and developing win-win solutions, helping the region build a diverse and sustainable economy. Harriet Bullitt does not oppose all logging. "Worse things can happen to the land," she notes, "such as a shopping mall."

Harriet Bullitt's newest enterprise is the environmentally friendly resort and retreat she created from the old CCC camp in Icicle Canyon. Besides being a vacation spot for individuals, it is also a conference center for small corporate, civic, educational, and environmental nonprofit groups needing a setting conducive to planning strategies, building teams, and resolving problems. "I had no experience in the hospitality business," she says, "but destiny brought this land into my lap. I had the money, I wasn't busy, and I had the time to put some thought into it. I wanted to provide a conference retreat where nature, the arts, outdoor recreation, and healthful dining inspire a reverence for Earth's life-giving wellspring, a place where people could come to restore their souls and to find something in their hearts to make the world a better place." She feels the woods and hills in Icicle Canyon evoke a sense of spirituality that she wants to share with others.

Bullitt well recalls the event that inspired her Sleeping Lady retreat. In 1991 she convened a meeting at Coppernotch lodge of fifteen salmon advocates who had been squabbling for years. From the ensuing dialogue, a coalition—Save Our Salmon (SOS)—evolved, a coalition dedicated to restoring habitat, filing lawsuits, educating the public on salmon and water-use issues, and promoting the idea that survival of the salmon meant jobs in both fishing and restoring spawning grounds. The Bullitt Foundation subsequently funded the coalition's work. Harriet Bullitt believes there is something about being in Icicle Canyon that influences group dynamics. Meetings in hotel rooms are dull and lacking in fun, she says. "Having some contact with the ground and the trees and the rocks and the river gets the cobwebs out of your head and helps thinking."

Called the Sleeping Lady after a mountain configuration nearby, the lodge encompasses sixty-seven acres and accommodates two hundred people in its forty buildings. It exemplifies the best in green renovation. The reconstruction process, for example, involved salvaging all usable wood from the existing cabins and other buildings. The remaining lumber and scrap drywall, after being pulverized on the site, ended up as mulch for trees and gardens. Ground-up scrap computer paper and cardboard boxes made up the ingredients for the cabins' insulation. As an alternative to cedar planking for decking, she used a new material called Trex, manufactured from scraps of hardwood and recycled plastic grocery bags. This prevented cutting thirty thousand linear feet of timber. Trex weathers to the same appearance as cedar.

The energy-efficient heating system recaptures and circulates heat from the laundry area and kitchen. The facility maintains an extensive composting center, fertilizing the organic garden that provides much of the retreat's food. Sleeping Lady's auditorium (seating two hundred), educational workshops, resident music group, library (with a wood stove), dance studio, swimming pool, and sauna and massage house enhance the

PHOTO: JENNIFER LIND

HARRIET BULLITT AT SLEEPING LADY LODGE, LEAVENWORTH, WASHINGTON.

getaway for ecotourists and groups year-round. A well-maintained system of cross-country ski trails abuts the lodge.

Bullitt invested over seven million dollars in the remodeling. Even though "green" renovation meant higher construction costs, she anticipates that the facility—which opened in the fall of 1995—will be profitable, producing additional funds for her philanthropy.

As president of Sleeping Lady, Inc., she relies on a full-time manager and a staff of over thirty persons, including a full-time horticulturist. She works at the lodge herself almost every day.

What gives Bullitt hope that humans will be able to save the planet from ecological ruin? As she sees it, "The reward is in the journey and in the worthwhileness of the goal. As long as the destination is unknown, there is always hope."

KATHRYN FULLER (1949-)
Shatters the "Green Ceiling"

Having lions on the Serengeti Plain chew on your tentlines in the middle of the night, Kathryn Fuller says, changes how you think about the world. She had accepted the invitation of two ecologists to accompany them to Tanzania for two months in 1973 to monitor wildebeest populations in the Ngorongoro Crater and study their behavior. This first direct experience with wildlife became the defining moment in Fuller's life. She never looked back.

Today, Fuller is president and chief executive officer of the World Wildlife Fund (WWF), which has been involved in forest protection throughout the world for decades. Since being promoted into the job early in 1989, Fuller, the first woman (and as of this writing still the only one) to serve as salaried head of a major U.S.-based conservation organization,[47] has doubled WWF's membership and revenue, which by 1996 had reached more than $72 million. The organization, now in its fourth decade and with 1.2 million member-donors, is part of a global network. It has over 50 branches or representatives covering 6 contingents, engaged in 3,000 projects in 140 countries. Fuller and WWF staff scientists are systematically identifying Earth's most distinctive and ecologically valuable regions so that protection efforts can be concentrated on them.

Focusing on rescuing species from extinction, conserving habitat, and reforming international markets that abet species and habitat extirpation, WWF under Fuller gives its highest priority to saving the earth's remaining forests. She points out that despite the high hopes and ideals of the Forest Principles Agreement adopted at the 1992 Earth Summit, deforestation rates have since increased. One of WWF's four major initiatives, Forests for Life, aims to reverse the degradation and destruction of forests worldwide. In September 1996, after twenty years of gathering data from more than eighty countries, and in collaboration with the

World Conservation Monitoring Center, the WWF released a unique world forest map, which showed that almost all of the globe's surviving forests lack protection.[48] Forest losses, whether involving quality or quantity, lead to soil erosion and hydrological disturbances, potential climate change, loss of species and genetic diversity, and hard times for indigenous forest dwellers.

The goal of the Forests for Life campaign is to persuade the world's governments to create protected forest areas that encompass at least 10 percent of what is left, to implement forest restoration, to practice sustainable logging, and to eliminate wasteful consumption by fostering changes in human behavior. Fuller believes that support for sustainable forest management is beginning to appear among forest product companies, consumers, and politicians.

⁂ For so long as she can remember, wildlife has entranced Kathryn Fuller. As a child growing up in Westchester County, New York, she devoured the books on animals and naturalists supplied by her conservation-minded mother. But later, intimidated by the pre-med orientation of Brown University's biology department, she switched her major to literature. After obtaining her B.A., she worked for a time at Harvard's Museum of Comparative Zoology. On the side, she sat in on Edward O. Wilson's biology classes, which renewed her enthusiasm for environmental science. She decided she wanted to become a conservation lawyer.

After graduating from the University of Texas Law School, Fuller sought a career in Washington, D.C. In 1977 she went to work in the Office of Legal Counsel in the U.S. Department of Justice. In 1979 she moved to the Land and Natural Resources Division to help create a wildlife law section, concurrently pursuing a degree in marine, estuarine, and environmental science at the University of Maryland.

Although she found her work heading the Wildlife and Marine Resources Section stimulating, she left the Justice Department when a third child arrived. (Fuller says that babies governed her career changes.) She then began working part-time as a consultant. One of her clients, the World Wildlife Fund, soon hired her to lead their program monitoring the international trade in endangered species. Fuller worked her way up the organization, functioning as general counsel, executive vice president, and director of public policy before moving into the top position. During this period she helped secure a ban on the worldwide ivory trade.

❧ The loss of tropical forests especially distresses Fuller. She says being in a forest makes her feel alive. Her commitment to changing the way we exploit tropical forests runs deep. In addition to protecting watersheds and being the world's largest pharmaceutical factory, these forests contain genetic material used to develop disease- and pest-resistant food crops as well as plant material containing natural insecticides. They also produce a host of marketable products apart from wood: nuts, fruit, shrubs, bark, ornamental plants, honey, and mushrooms, to list a few examples. Tropical forests are home to more than half (some experts say 90 percent) of the earth's plants, mammals, birds and insects.[49] On a single tree in Peru, the biologist and ant expert Edward O. Wilson found forty-three species of ants (the same number as found in all of Britain).[50] If the current rate of destruction continues, part way through the next century tropical forests will be gone, and with them the benefits to humanity and the world of a host of as yet undiscovered life forms.[51]

Most tropical forests, circling the equator, lie in the poorest countries, which must struggle to pay even the interest on loans from foreign lending institutions. One strategy that Fuller employs at WWF to preserve tropical forests is conservation finance—promoting trust funds and debt-for-nature swaps. Tom Lovejoy, then a WWF staff ecologist (now with the Smithsonian Institution), brought attention to the debt-for-nature idea in an op-ed piece he wrote for the *New York Times* in 1984. The process can be complicated and involve intricate negotiations, but simply stated it works like this: A conservation organization acquires a portion of the developing country's foreign debt, either by donation or purchase, at a discount from its face value (resulting from the lender's fears of never being repaid at all). The conservation organization then offers to pay off the debt, provided the developing country agrees to dedicate additional resources of its own to local conservation programs.

For example, a WWF debt-for-nature swap of $4.5 million in Madagascar has, since its inception in 1990, made it possible for the country's Waters and Forests Department to train nearly four hundred wildlife protection deputies who work with local communities to safeguard their forests. The communities have since planted two million trees and created almost a thousand tree nurseries; several hundred local environmental groups have sprung up as well.

Since 1984, the World Wildlife Fund, Conservation International, and the Nature Conservancy have all negotiated such swaps. The transactions have so far eliminated two hundred million dollars (U.S.) of the debt of developing countries in Africa, Asia, and Latin America,[52] at the same time providing substantial new funding for local conservation projects. Fuller contends that making conservation one of the chips on the table when national debts are renegotiated is the next step. For too long the silent—and unpaid—partner in economic development has been the environment.

Fuller believes protecting and preserving the natural environment in both industrialized and developing countries ranks in importance with maintaining peace and political stability. But, she says, "Conservation only works if grounded in local support and initiative. My biggest challenge is figuring out how to foster that. . . . As the world has become more complex and conservation problems more serious, our work is increasingly characterized by an emphasis on the human components of the nature-conservation equation."[53]

Nepal is a case in point. The heart of the Himalayas and home to Mount Everest, it attracts as many as twenty-five thousand tourists each year. Even though they provide much-needed revenue for the cash-poor Nepalese, the trekkers also bring scores of problems with them, the primary one being the need for firewood. One mountain trekker, plus guides and porters, consumes ten times more wood each day than does one Nepalese village family.

Seeing that the wood consumption rate was exceeding the sustaining capacity of local forests, which provide habitat for several endangered species, WWF in 1985 pioneered a new type of community-based protected area in the Annapurna region of Nepal that links conservation and human needs. One American and two Nepalese conservationists, hired by WWF, engaged the local villagers and community leaders in several months of discussion to evaluate the region's needs and develop a program proposal. The Annapurna Conservation Area project emerged, directed by a Nepalese conservationist. The community now manages its own natural resources, with local conservation committees making the decisions. In the ten years following its inception, the project has become self-sufficient through tourist revenue. Wood plantations have relieved the burden on the forest. Trekkers must buy cooking kerosene from village-operated concessions. Forests in the Annapurna Conservation Area are regenerating, restoring the natural habitat for the endangered species living there—red pandas, musk deer, and snow leopards—as well as for rare bird and plant species. Fuller comments, "We know that innovation is a critical component of lasting conservation success. We know that, at the end of the day, individual action is what makes conservation work."[54]

She says that the most difficult part of her job is setting priorities. Although WWF is a large organization, its resources are limited relative to the number of pressing global problems. Another obstacle Fuller encounters involves public attitudes. "Many of the issues we work on are considered important, but not critical, by the public . . . not as important

PHOTO: MARTHA VAN DER VOORT, WORLD WILDLIFE FUND

KATHRYN FULLER DISCUSSES THE PROGRESS OF THE ANNAPURNA CONSERVATION AREA IN NEPAL WITH THE PROJECT'S DIRECTOR, MINGMA NORBU SHERPA.

as health and economic issues. One of the toughest challenges is raising awareness sufficiently so that people—both decision makers and the general public—embrace conservation as a priority objective."

That is why, of the multitude of WWF's diverse projects around the world, Fuller ranks one as urgent for saving the planet: an international environmental education program. In her view, more conservation education is needed to maintain an informed and passionate constituency in a world where short-term interests dominate. To that end, she has initiated Windows on the Wild (WOW), an educational program using biodiversity as its organizing theme. Through a full-color magazine primer, teacher-training programs, special events, and an interactive Internet Web site, WOW provides an entertaining approach to helping both young people and the general adult population better understand biodiversity, thereby engendering a more environmentally literate citizenry. Fuller and WWF collaborated with a number of other educational and environmental organizations as well as government agencies in developing WOW's array of teaching tools, which are now being incorporated in middle-school curricula around the country.

Fuller takes her greatest pleasure from being in the field—whether walking in a forest, scuba diving, or watching thousands of animals migrate across the African savanna. "I derive real satisfaction from getting my hands dirty once in awhile—getting something running, being on the ground and able, by my personal effort, to have made a tangible difference someplace." These experiences re-energize her, she says, to deal with the demands of day-to-day management.

Does her way of managing differ from that of men? "It may be somewhat more inclusive—a very open management style. . . . I put heavy emphasis on giving people as much responsibility as possible and letting them run and get stuff done. It is a decentralized approach. Part of my philosophy is to find the best people you can and give them the resources and the running room they need within a clear set of priorities—and then not get in their way." Fuller claims that managing a household offered some of her best training. The priority-setting, listening, and organizing skills required, she says, are the same ones needed in an institutional setting.

What keeps her going when she encounters disappointments or setbacks in her efforts? "Being able to see our track record of success over more than thirty-five years, and knowing that if you just keep at it, more successes will come. This, coupled with confidence that what we are doing is terribly important and urgent—a mission—and that we have been able to overcome obstacles in the past."

A recipient of the United Nations Environment Programme's Global 500 award, Fuller also has the distinction of having a species of ant named after her. She once asked ant expert and WWF board member Edward O. Wilson to look at the ants on her desk. After collecting them and completing taxonomic work back in his laboratory, Wilson discovered that they were a new species. So he named them after her.

What gives Fuller hope that humans will be able to save the planet from ecological ruin before it is too late?

> I derive hope from seeing how much further along we are in gaining access to decision makers in government and in the corporate world. Many of these institutions, though they may not be working perfectly, have put in place policies and procedures that help protect the environment. Just seeing that much transformation in a decade certainly helps give one hope that they have recognized the need to act and that the problems really do exist—and that humans will be able to mobilize and come up with solutions. It is not easy, and there is much resistance. But we have seen quite a dramatic transformation in access across the board.

Reflecting on what she hopes to accomplish during her tenure as president of WWF, Fuller says, "I'd say it would be to see tangible improvements in the status of key species—for example, pandas and tigers—and in key habitats such as the Amazon forests, the Florida Everglades, as well as coral reefs."

CHAPTER TWO

*[W]omen, out of their own culture, can initiate action to
change their life situation and the nature of society itself.*

ELISE BOULDING
The Underside of History

First Lady of Environmental Science

ELLEN SWALLOW (1842-1911)

Late in 1870 Ellen Swallow, then twenty-eight, wrote to Dr. J. D.
Runkle, president of the Massachusetts Institute of Technology, seeking
admission. No woman had yet been accepted at the five-year-old school,
and her friends had told Swallow it would be foolish even to try.

Women confronted obstacles in obtaining an education in the 1800s.
Medical men spread the belief that learning would adversely affect
women mentally, physically, and emotionally. Swallow, the only child of
schoolteacher parents, spent her early years on their farm near Dunstable,
Massachusetts. Her frail health and the poor quality of the local school
prompted her parents to educate her at home.

For a bright youngster like Swallow, home schooling stimulated
hands-on, experiential learning, inquiry, observation, and exploration.
Cultivating her own garden, exploring the woods and streams nearby,
collecting and classifying rocks, plants, and fossils, writing about her observations, and reading avidly on her own provided creative and self-
motivating circumstances for her early education. Not until she was six-

teen, when her family moved from their farm to the nearby village of Westford, did Swallow attend a community school.

In addition to studying classical literature, Latin, and French, Swallow demonstrated a special talent for mathematics at the Westford Academy. She soon began tutoring other students. Swallow coached her classmates from her own first-hand experiences and observations: when it came to nature, her knowledge usually exceeded the teacher's. Working after school with her father in his general store expanded her education to include administrative skills and business management.

Now twenty-six, Swallow yearned for higher education, but her mother's chronic illness and a lack of money frustrated her. Besides, no colleges for women existed in New England at that time. However, upon learning about Vassar, the new college for women that had opened in Poughkeepsie, New York, in 1865, she immediately applied and passed the preliminary entrance examinations.

Swallow did so well that she became a senior in her second year. Two particular members of the Vassar faculty had a great influence on her— the world-renowned astronomer Maria Mitchell, who had discovered a comet in 1847, and Professor A. C. Farrar, who headed the Natural Sciences and Mathematics Department. Through the Vassar Observatory's telescope, Swallow discovered star clusters and nebulae that Maria Mitchell could not identify. And when the Smithsonian Institution requested someone to keep the meteorological record at Vassar—and supplied the instruments—Maria Mitchell chose Swallow for the job. As a result of this experience, forecasting the weather became one of her lifelong hobbies.

Torn between the disciplines of chemistry and astronomy, Swallow eventually decided to major in chemistry. The practical applications of such knowledge intrigued her, whereas the cosmos seemed far removed from everyday experience. She registered for almost all the science and math classes offered at Vassar. Although she studied animal and plant life, soil, rocks, and fossils, observing water especially fascinated her. During her months at Vassar she supported herself, barely, by tutoring other students in Latin, math, and German for $1.50 a day. The college granted her a small scholarship during her last year.

After she received her bachelor of arts degree, Swallow wrote to several chemical companies seeking an apprenticeship. None of them hired women. But one, Merrick & Gray in Boston, suggested that she approach the city's new institute of technology—remarkable advice to a woman in 1870.

In her letter to the Massachusetts Institute of Technology she asked if it accepted women. She named Maria Mitchell and Professor Farrar of the Vassar faculty as references. In December 1870, a few days after her twenty-eighth birthday, Swallow received a letter from President Runkle—a letter of historic significance for women's education—advising her that she could attend MIT without charge as a "special student."

Swallow could hardly believe the school was going to allow her to enroll without paying tuition. She assumed her poverty had motivated the trustees' decision. Many years later, she learned the real reason: if trustees or students complained about her presence, Runkle could claim nonstudent status for her. She said if she had known that at the time, she would not have enrolled.

Realizing that many disapproving eyes were watching the "Swallow experiment," she kept a low profile, taking care to perform extra tasks like sweeping floors, dusting, washing beakers, and mending professors' suspenders or sewing on their buttons that were considered "women's work."

Swallow's outstanding academic performance and her analytical skills in the laboratory gained the attention of faculty members. One of them, Professor William Ripley Nichols, who initially did not believe in women's education, made her his assistant during her second year at MIT, assigning her the responsibility for supervising a major project for the newly established Massachusetts Board of Health. The project involved a sewage and water supply survey throughout the commonwealth that took two years to complete. In his report to the board in 1874, Professor Nichols pointed out: "Most of the analytical work has been performed by Miss Ellen Swallow, A.M. [Vassar, 1873], in the laboratory of the Massachusetts Institute of Technology, under my direction. I take pleasure in acknowledging my indebtedness to her valuable assistance and expressing my confidence in the accuracy of the results obtained."[1] Because of her role in this pioneering study, Swallow became an internationally recognized water scientist even before she graduated from the institute.

In 1887 she supervised another MIT project for the Massachusetts Board of Health, an undertaking characterized by her biographers as Swallow's greatest contribution to public health. The work required collecting forty thousand water samples from all over the commonwealth, analyzing them promptly, and recording the results. Water analysis emerged as a new branch of chemistry at that time, and MIT's Sanitary Chemistry Lab shone as the first of its kind in the world. Water analysis became Ellen Swallow's focus for a decade. She considered clean drinking water to be a basic human right. In 1911 Swallow remarked: "Water [is] a national asset. Preservation of *quality* of water for man's uses is a public duty. . . . The very life of the nation depends on its water supply. . . . The time is here, already come, when the preservation of the quality and quantity of such water as remains to us is of paramount importance."[2]

ELLEN SWALLOW TESTING THE WATER IN JAMAICA POND.

Even earlier, she had observed: "In common law, water is held to be a gift of nature to man for use by all, and therefore not to be diverted from its natural channels for the pleasure or profit of any one to the exclusion of the rest. Neither has one the right to return to the channel water unfit for the use of his neighbor farther down the stream. That is, there is no private ownership in surface waters flowing in natural channels."[3]

Swallow's water survey for the Massachusetts Board of Health resulted in the world's first water purity tables and enabled the commonwealth to create the first water-quality standards in the nation. It also led to the establishment of the world's first modern sewage treatment systems and helped launch the public health movement in the United States.

Swallow traveled extensively in North America and abroad throughout her life, lecturing, consulting, teaching, and attending association meetings. She seldom left home without her portable laboratory for gathering and analyzing samples of water, and she tested water from Jamaica to Alaska. In 1903 she commented that there was hardly a place in the world where water did not show the effect of contamination by humans.

In the last quarter of the nineteenth century, the environment in the urban Northeast suffered from the high concentration of American industry in the region. Often built along waterways, industrial facilities disposed of their soluble or suspendable wastes by dumping them in rivers; they tossed their rubbish, garbage, slag, ashes, and scrap metal indiscriminately on land. The widespread use of coal, with its devastating smoke, as the primary fuel polluted the air; soot settled on buildings, laundry, and in people's lungs. Raw sewage spewed into rivers and streams. The use of horsedrawn vehicles presented severe waste disposal problems, since a single horse could create gallons of urine and almost twenty pounds of manure a day, all discharged in the streets. The total clearing of trees and plants from the cities to make way for commerce, dwellings, and streets robbed urban areas of oxygen generated by the trees. Food transported from farms to cities in uncovered, unsanitary vehicles without refrigeration introduced poisons at the dinner table.[4] The meat-packing industry often sited slaughterhouses and associated tanneries and glue factories in residential areas. Animal wastes piled up on vacant lots, tanneries washed hides in city watering places, and foul stenches drifted through neighborhoods. Added to all this was the noise level of urban factories, which in many areas approached a constant roar.

Pervasive ignorance reigned about the cause and prevention of diseases like pneumonia and tuberculosis, which swept through the cities in the winter; like malaria and typhoid fever, which claimed lives in the summer; and like diphtheria, smallpox, scarlet fever, and cholera, which struck at any time. The death rate in Massachusetts for children between birth and nineteen in 1885 reached 300 per 1,000.[5] Water, so necessary to all life, could also be a vehicle of death, as the scourges of cholera and typhoid fever demonstrated.[6]

Swallow specialized in chemistry because of her desire to apply its principles to the severe environmental problems she saw all around her. She believed chemistry should be dedicated to public health. In her book *Conservation by Sanitation,* she wrote: "Air and water are two of the conditions now most before the world. . . . Man has made himself much extra trouble by reckless waste of nature's provision."[7]

The first woman to receive a science degree from MIT, Swallow also received in the same year (1873) a master's degree from Vassar for her groundbreaking thesis on the rare metallic element vanadium. She had become interested in mineralogy through her association with Robert Richards, an MIT faculty member. Richards, too, had been opposed to coeducation, but like his colleague Professor Nichols he reversed his views. His change of heart came when Ellen Swallow, fluent in German, translated the texts of his German professional journals on mineralogy for him.

Immediately upon graduation, Swallow, then thirty-one years old, applied for membership in the American Association for the Advancement of Science (AAAS). She wanted a doctorate in chemistry from MIT and continued her graduate study toward that goal for two years. However, the department's faculty did not like the idea of a woman being awarded its first doctoral degree. The AAAS, on the other hand, recognizing her level of scholarship, elevated her to the status of Fellow four years after she joined, such designation being reserved for scientists performing above the doctoral level. And in 1879 she became the first woman member of the American Institute of Mining Engineers, a society of four thousand men who closely guarded the professional standards they had established for membership.

The only doctorate Ellen Swallow received came in 1910, a few months before she died, when Smith College conferred on her an honorary doctor of science degree. The Massachusetts Institute of Technology, where she taught for over a quarter of a century, finally recognized her with a bronze plaque, installed in 1924, in a hall of the chemistry building.

When Swallow graduated from MIT in 1873, the institute had indicated she could stay on if she wished, without pay, as a "resident graduate." For income she had developed a large private consulting practice in sanitary chemistry, testing not only water but also air and food. On the recommendation of Professor John M. Ordway at MIT, who maintained a large industrial-consulting practice, the president of the Manufacturers Mutual Fire Insurance Company, Edward Atkinson, in 1884 appointed her the firm's consulting chemist. In this capacity she pioneered research on the danger of spontaneous combustion posed by various oils used in textile mills, where men, women, and children faced constant threats of dismemberment and death from fires and explosions. Most of these fires resulted from friction in the machines and spontaneous combustion, a situation exacerbated by the mill owners' use of cheap, low-grade oil to cut costs.

Swallow designed and supervised experiments to test the oxidation of various oils and dyes that came into contact with textiles and other inflammable materials in the mills. Her analysis gave the insurance company the data necessary to force suppliers to upgrade their oils and to force mill owners to raise their safety standards and redesign their factory procedures. Her evaporation test for volatile oils became a world standard.

Five years after she graduated, MIT made Swallow an assistant instructor, but still she received no pay. Ten years later, her title was upgraded to instructor of sanitary chemistry. Thereafter, she received no promotion, increase in salary, or change in title. Yet she helped inaugurate at MIT the world's first comprehensive sanitary engineering course and taught chemistry and sanitary engineering to countless students, mostly men. She also taught courses in the analysis of air, water, and food and the chemistry of water and sewage. In addition to teaching, she functioned as dean of women at MIT, although without that title.

In the Kidder Laboratories at MIT, Swallow, using simple language, employed a holistic method of training sanitary engineering students, drawing from all the sciences—an interdisciplinary approach that even today is not commonly practiced among scientists.

Swallow joined the teaching staff of the Society to Encourage Studies at Home and developed a new science correspondence course for intellectually isolated women, writing textbooks for them, corresponding with them, and even sending them microscopes, other laboratory materials, and minerals. In one of her annual reports for this project, she remarked: "We aim to unclasp for our students the book of nature. . . . They are mentally starved to death."[8] She taught the home students geology, mineralogy, and physical geography, and supervised the teaching of botany and math. They also received consumer information on the basic physics of home heating and ventilating and on how to site a home on a lot for proper drainage and sanitation. Swallow later modified the course to meet the needs of college-level women students until greater opportunities for enrolling formally became available to them.

Ellen Swallow was already twenty-eight when she entered MIT, long past the age women of the day customarily married. Her observations of the marriage relationships of friends and acquaintances had turned her away from any interest in the institution. But during her second winter in Boston, Robert Richards, a professor of mining engineering and head of MIT's new metallurgy laboratory, quietly began to court her. The same month in 1873 that Swallow received her science degree, Professor Richards whispered his proposal to her—in the MIT chemistry laboratory. But Swallow told him that she needed time to think the idea over.

Ellen Swallow and Robert Richards were opposites in personality and temperament. A dynamic woman with inexhaustible energy, she was quick both mentally and physically, whereas Richards was unhurried and deliberate. But their common interest in scientific pursuits sparked an immediate and abiding bond.

It took Swallow two years to make up her mind. She refused to be influenced by societal attitudes. She believed women need not lose their individuality in marrying—that they have a personality not under a husband's control and that simply being married did not make them devoted slaves. Fortunately, Richards demonstrated a parallel eagerness to pioneer a new type of marriage.

Early in June 1875 they married. The couple honeymooned by taking a field trip to Nova Scotia, accompanied by Richards's entire mining engineering class. For the outing, Swallow wore a short skirt and heavy boots, a practical but shocking costume to her friends. In an age when women cinched their waists with tightly laced corsets, Swallow refused to conform. She once wrote to a friend that clothes should be a comfort and a protection rather than a burden, and that if removing one's clothes at night felt like a big relief something was wrong.

The environment that Swallow and Richards created in their book-filled, three-story home in Jamaica Plain—at that time four miles outside the city limits of Boston—demonstrated every aspect of their knowledge of science as it related to human health: clean air and water, sanitation and proper sewage disposal, cleanliness and hygiene. To rid the dwelling of stale air and toxic fumes, they installed windows that opened at both top and bottom for ventilation in a coal-heated dwelling; in addition, a master skylight vent, with exhaust fans, in the third-floor hall remained open most of the time. Swallow switched from coal to a gas stove for cooking. As soon as electricity became available, she made that conversion for lighting, eliminating the fumes and flames of gas.

Since Jamaica Plain lacked city water and sewer systems, Swallow analyzed water from an old well she discovered under her porch; she found it to be uncontaminated. And she had the sewage drain pipes lengthened to take them farther from the well. She oversaw the renovation of the indoor plumbing, replacing the old lead pipes. In other words, in the 1880s Swallow designed a prototype for a healthful home employing scientific principles for ensuring clean water, clean air, and a sanitary waste management system—the same principles she promoted for hospitals, schools, and factories.

Her decor departed radically from what was popular at the time. Instead of thick carpeting that was difficult to clean, she chose polished hardwood floors and small area rugs. Instead of heavy, dust-collecting drapes at the windows, she dressed her window areas with plants and flowers. "Where plants will not grow people ought not to live, is a safe maxim," she said.[9]

She created in her kitchen the nation's first testing laboratory for consumer products, christening it the Center for Right Living. She picked a few women students from MIT to assist her with testing. In exchange, they received free room and board in the Richardses' home.

From the time she first entered MIT, Swallow had a vision of a science lab for women at the institute. When she married and personal financial pressures lessened, she focused most of her energy on making it happen. Remembering the earlier willingness of the Women's Education Association of Boston (on whose board she served) to help raise money for a chemistry course she had taught at a Boston girls' high school while a senior at MIT, Swallow resolved to appeal to them again. Several months after her marriage, she presented her case to the association in a moving speech about the educational needs of women in the sciences. The association agreed to help. She had already persuaded MIT to designate space for a women's lab—an old garage that had been slated for renovation into a men's gym. Swallow not only served as chief fund-raiser for the women's lab, helping as well to pay for some of the initial equipment out of her own pocket, but she contributed $1,000 each year for operations, new equipment, and scholarships.

Though still without salary from MIT, Swallow accepted the assignment of being in charge of the day-to-day operation of the world's first women's science laboratory and helped instruct the students. The courses included chemical analysis, industrial chemistry, mineralogy, and chemistry as related to vegetable and animal physiology.

Swallow wrote a textbook for schoolteachers to use in educating children about the environment. In this book, *Sanitation in Daily Life*, she declared: "Human ecology is the study of the surroundings of human beings and the effects they produce on the lives of men. The features of the environment are natural, as climate, and artificial, produced by human activity, as noise, dust, poisonous vapors, vitiated air, dirty water and unclean food."[10]

Her work encompassed not only the testing of food, water, and consumer products but of air also. "The three essentials for healthful life," she wrote, "are food, water and air. In this most medical men and hygienists are agreed. While we eat perhaps three times a day and take water every few hours, we breathe, upon the average, twenty times a minute, or 28,800 times every 24 hours. So constant a function must be an important one."[11] Swallow, a member of the American Public Health Association, served on its Committee on Standard Methods for the Examination of Air.

Seven years after the first women's lab at MIT opened, the institute tore down the makeshift quarters and moved the lab into the new chemistry building. Swallow organized a fund-raising effort for a women's lounge in the new women's lab—a feature not included in the institute's plans. Finally, thirteen years after Ellen Swallow had enrolled in MIT and paved the way, the "special student" label for women disappeared.

Swallow's personal crusade for the education of women in environmental science reached even elementary schools. She believed a beginning must be made in childhood if women were ever to attain control over the conditions of their lives. She therefore introduced science instruction in Boston's public elementary schools. She created and taught a mineralogy course and wrote an accompanying pamphlet, *First Lessons in Minerals.*

Her views on effective education stemmed from her own early home-schooling experience. She believed education should be broader than the three R's and should become more free. It should include exploring our environment—finding out what it is and how to live in it. Hands-on, experiential learning, she believed, stimulated interest, observation, and creativity in students. Children need to be actively doing things, she said, and requiring five-year-olds to sit still even for an hour amounted to cruelty.

While teaching this elementary school course, Swallow conducted an experiment to compare her teaching methods with conventional ones. She gave an identical set of lessons to the schoolchildren and to a class of Harvard undergraduates. Instead of consulting textbooks for their conclusions, the youngsters trusted their own observations and identified and classified minerals more quickly than did the Harvard men.

Women's groups in every era function as catalysts for change. The organization known today as the American Association of University Women (AAUW), a group that has made higher education possible for millions of women, began life as the Association of Collegiate Alumnae. Swallow presided at the organizing meeting in mid-January 1882. Once the association had been launched, she backed away from accepting a conspicuous leadership position, but agreed to serve on the executive committee and to lead various committees created for specific projects.

Soon local branches of the association sprang up. In the Boston group, Swallow organized a sanitary science club, whose members conducted environmental quality and efficiency audits of homes and institutions in the city. She also worked with the Boston branch to bring about an investigation of sanitary conditions in the public schools.

In 1896, at the annual meeting of the American Public Health Association in Boston, Swallow charged politicians, parents, and the general public with the deaths of at least two hundred schoolchildren each year. She said half the schools in Boston were unhealthy, with open sewer pipes, filthy toilets, dirty floors, and lack of ventilation. These illegal and unsanitary conditions caused more than five thousand cases of illness per year. Only a few buildings had working fire escapes. Boston teachers suffered as well, having the highest death rate in the nation.[12]

When the city's politicians refused to budge, Swallow appealed to a wider public, writing articles for national periodicals and giving speeches. Finally, she and her special committee of the Boston branch of the Association of Collegiate Alumnae went over the heads of the local politicians and appealed to the Massachusetts legislature. Although the legislation proposed by Swallow's committee did not pass in its original form, a watered-down version became law. At first, it seemed like a defeat; several years later, however, additional legislation transformed school conditions. The ultimate overhaul extended beyond sanitary reform to include all aspects of schooling in Massachusetts—administration, curricula, personnel, and the responsibility of the entire community for education.[13]

In the 1800s adulterated food products commonly reached the family dinner table. No pure food laws existed. Always the crusading pioneer, Swallow, along with her students at the new women's lab at MIT and in her Jamaica Plain home laboratory, began the task of testing staple foods and other household products for the Commonwealth of Massachusetts in 1878–79. They found, for example, that some samples of cinnamon contained no cinnamon at all, only mahogany sawdust. Under the direction of the New York State Board of Health in 1882, their examination of pepper showed that 70 percent of the commercial product was adulterated.[14] Swallow and her students found starch and alum in baking powder, starch in mustard, alum in bread, sodium chloride and sand in sugar, and acid phosphate of lime passing as cream of tartar. They discovered watered-down milk and tainted meat disguised in sauces. After the Massachusetts Board of Health published Swallow's report in 1879, the commonwealth between 1882 and 1884, using her analyses, passed the first of its Food and Drug Acts.[15] The work of testing consumer products extended to papers, furniture, fabrics, wood, and appliances. Among other things, she found arsenic in wallpaper and mercury in fabrics.

The 1 December 1892, issue of the *Boston Daily Globe* carried an article headlined "New Science: Mrs. Richards Names It Oekology." The previous evening, before a crowd of three hundred distinguished businesspeople and a few scientists in the banquet room of Boston's Vendome Hotel, Ellen Swallow Richards introduced the term "oekology"—later modified to "ecology"—to Americans. In her view, "Ekology is the worthiest of all the applied sciences which teaches the principles on which to found healthy and happy homes."[16]

She did not coin the word. In 1873 Ernst Haeckel, a biologist at the University of Jena, who had already named a number of new sciences, came up with a name for the study of organisms in their environment—"oekologie." Being fluent in German, Swallow researched the genesis of the word. *Oik* is derived from the classical Greek word for "house"; *oek* represents "every man's house," or our universal environment. As a pioneering interdisciplinary scientist, Swallow embraced the most comprehensive view of what ecology, or environmental science, should include. In her extensive teaching, lecturing, writing, and consulting work over the years, she conjoined the life sciences, physical sciences, earth sciences—and even the social sciences. In her concept of ecology, all the branches of science overlapped. And the human factor—improving the environment for people—could not be ignored.

But science purists of the period considered this philosophy subversive. To them, ecology had a narrow definition relating only to plant and animal life. Swallow's colleagues criticized her, in fact, for not engaging in more pure-science research. She, on the other hand, failed to understand why scientists took so long to apply the results of their laboratory research to everyday problems. When she felt the pressure of social change, she said, research had to be put aside.

With her vision of the scope of ecology rejected, Swallow took another tack to bring her own brand of environmental science to the nation's homes, schools, and industries. She believed that, in self-defense, home managers needed to have some knowledge of chemistry, nutrition, the health effects of air and water quality, and waste management. Eventually, the American Home Economics Association came into being, and she served as its first president. Although Ellen Swallow did not actually found the home economics movement, as some reference sources aver, she recognized a niche in this existing, but fragmented, movement that allowed her to introduce her vision of how environmental science could be applied in homes and schools.

She cautiously began substituting the term "domestic science" for "ecology" in her lectures and writings. In 1897 a national opinion magazine, the *Outlook*, published an article by Swallow in which she defined domestic science as an interdisciplinary applied science. If the purists would not accept her "subversive science," why not create a new interdisciplinary profession—a sort of "home ecology"?

Lending assistance to her new effort, her friend Melvil Dewey (creator of the Dewey decimal system) helped her to get domestic science into the New York Regents exams in 1896. At the time, he was director of the New York State Library and one of the regents controlling the state's higher education systems. From his home at Lake Placid, he counseled her to separate herself from the life sciences community and to pursue an independent path with her interdisciplinary knowledge, which had already aroused widespread interest among thousands of people.

Melvil Dewey offered his Adirondack retreat as a regular meeting place for Swallow and her growing corps of coworkers. In 1899 she led the first meeting of what they called the Lake Placid Conference. Specialists in chemistry, biology, economics, psychology, and sociology participated. Eleven years later, the movement had become a nationwide coalition of related groups, and it adopted a new name: the American Home Economics Association. Swallow's converting of her concept of ecology, or domestic science, to a social science afforded many more people access to it than would have been the case had it stayed in the private domain of the life scientists.

Swallow launched a professional journal, the *American Kitchen Magazine*, to reach the association's field-workers. The publication, which was very well received, carried scientific papers from other countries and featured the home science of a different nation each month. One article explored the valuable but overlooked insights of the North Dakota Indians on environmental management. Swallow set up a company to publish the magazine, the Home Science Publishing Company.

She succeeded in getting environmental science courses into universities across the nation, as well as reaching professional and trade schools, elementary and high schools, and the sponsors of correspondence and extension courses. The movement soon became part of the U.S. Department of Agriculture. It grew into an international endeavor when it spread to England, Canada, and Australia. But over the years the home economics movement lost the strong environmental science slant that Swallow had originally envisioned.

In 1894 the citizens of Poughkeepsie complained to Vassar about the college's discharging of its raw sewage into Casperkill Creek, the source of their drinking water. Swallow, a newly elected trustee, listened to the college's proposed solution: at a cost estimated to be between $37,000 and $50,000, a six-mile sewer line could be laid to carry the raw sewage directly to the Hudson River, bypassing Casperkill Creek. Swallow considered discharging raw sewage into public waters a medieval practice, no matter whether its destination was the Casperkill Creek, the Hudson River, or the local reservoir. Offering her own alternative, she sketched a picture on the blackboard for the trustees of a new and reliable system for treating waste, a sewage treatment plant, which would not pollute any rivers or make anyone sick. The remaining sludge, she pointed out, could be used as fertilizer for local crops. When the trustees, assuming it would be prohibitively expensive, asked the cost, Swallow assured them that the plant would cost only $7,500.

Ellen Swallow died penniless at age sixty-eight, but not because she had failed financially. She simply kept giving her money away. During the last few weeks of her life, in March 1911, even though suffering from angina pectoris, she used her fading energy to write an address she had been asked to present at MIT's fiftieth anniversary celebration, a Congress of Technology. Her husband, Professor Robert Richards, hand-carried the speech to MIT's president a few days before her death. The address was entitled "The Elevation of Applied Science to an Equal Rank with the So-Called Learned Professions."

On the day the Boston papers reported her funeral, news broke of the indictment of five companies for violating the new food and drug laws. Also on that day, the papers carried a story urging enforcement of new city ordinances to make Boston's schools safe and clean.

In 1912 MIT received a gift of $15,000 to establish the Ellen Richards Memorial Fund to help deserving sanitary scientists and engineers finance their education. Scholarships from this fund are still awarded today. Another major gift to MIT in 1961 made possible the creation of an Ellen H. Richards Professorship, which is still filled. The bronze plaque of Swallow's image—unveiled in 1924—is now in the Ellen H. Richards Lobby of the Lavoisier Chemistry Building at MIT, along with a wall-case display of photographs and a painted portrait.

Nine years after Ellen Swallow died, the AAUW raised over $156,000 to enable Marie Curie to purchase the one gram of radium she needed to continue the groundbreaking research that led to her second Nobel Prize. When Curie made her first trip to the United States, some years after Swallow's death, her only formal address to an American audience was at Vassar. She titled it "An Ellen Richards Monograph."

The citation accompanying Swallow's honorary doctor of science degree from Smith College sums up her major achievements.

> Ellen Henrietta Richards, Bachelor and Master of Arts of Vassar College, Bachelor of Science of the Massachusetts Institute of Technology, and there for over a quarter of a century instructor in Sanitary Chemistry: by investigations into the explosive properties of oils and in the analysis of water, and by expert knowledge relating to air, food, water, sanitation, and the cost of food and shelter, set forth in numerous publications and addresses, she has largely contributed to promote in the community the serviceable arts of safe, healthful, and economic living.[17]

In her day, Swallow bore the brand of "reformer"—a term not meant to be complimentary, since most reformers were women. She criticized medicine and made enemies in industry, government, and labor. She offended city planners and slum landlords. And she egged citizens on to take action, urging them to become familiar with local laws and to report infringements to the local health department. She pressed people to demand that cities clean up their air, improve trash handling, and establish sewage treatment systems. And all this almost a century ago.

CHAPTER THREE

*The support system for women found in women's
organizations and networks acts as a great multiplier of
individual effort.*

ELISE BOULDING
The Underside of History

Early Municipal Housekeepers

Although the nineteenth century witnessed the freeing of serfs and slaves, women's roles and status regressed in the Victorian Age, robbing them of self-confidence and any shred of independence.

Nevertheless, women in the nineteenth and early twentieth centuries empowered themselves by creating their own societal niches. Organizing themselves into clubs, they established powerful bases for citizen action. The discussions and pursuits of these collegial groups also provided many women with a substitute for the formal higher education often unavailable to them.

The rapid urbanization and industrialization characteristic of this period introduced severe environmental pollution. Concern for the health and safety of their families fired women's groups to undertake sanitary-reform campaigns in their communities. In cities all over the country, women's groups strove to clean up streets, inspect food markets, improve air and water quality, and reform waste disposal practices.

Women's groups acquired the reputation of being radical because women had little tolerance for slow-moving and indecisive bureaucracies. For example, trash containers on city street corners today are ubiquitous. The idea originated in Philadelphia with a women's group in the mid-

1890s. After receiving approval from city officials, members of the Women's Civic Club of Philadelphia purchased trash barrels with their own money and placed them in selected areas of the city. Later, the club donated the containers to the city council. The resulting improvement in the cleanliness of areas having refuse baskets inspired the council to purchase more containers and expand the coverage. About two decades later, in 1913, Edith Pierce became the first woman appointed to the post of street-cleaning inspector of Philadelphia.[1]

CAROLINE BARTLETT (1858–1935)
Sanitizes Cities

Around the turn of the century, a college-educated Unitarian minister and social reformer in Kalamazoo, Michigan, found herself catapulted into a second career as a nationally recognized sanitation expert. And it happened almost by chance. As leader of a women's study group in her church, Caroline Bartlett had been casting about for a speaker on meat inspection. When all the officials she had invited declined, she decided to give the talk herself. In preparation, she and several other clubwomen toured the seven slaughterhouses selling meat to Kalamazoo's markets. To their horror, the women discovered the worst possible sanitary conditions: buildings in disrepair and decay; surfaces in the slaughterhouses covered with thick layers of grime, grease, mold, and hair, all caked in blood; and slaughter of diseased along with healthy animals.

When these conditions were disclosed in the local newspaper and at a city council meeting, the community was shocked into action. However, because the slaughterhouses were located outside the city limits, the council found it had no jurisdiction over them. Caroline Bartlett, married to a doctor, took the case to Michigan's Board of Health. When it failed to act, she surveyed meat inspection laws in other states, drafted a bill giving Michigan cities authority to institute meat inspection regulations, and, after the bill's introduction, lobbied the members of the state legislature in Lansing in its support. In 1903 the bill passed. Bartlett then shepherded a model local ordinance through the Kalamazoo city council.

She went on to create a Women's Civic Improvement League in

Kalamazoo. Under Bartlett's leadership, this group studied efficient street-cleaning methods elsewhere and conducted a three-month demonstration in Kalamazoo of a new, more effective, and less expensive system. She even had the street cleaners wear white uniforms. She and other club members won the cooperation of merchants, persuading them to clean the streets and walkways around their stores. She publicized unflattering photos of substandard conditions when necessary. The city council finally put the recommendations of the women's league into practice, but only after an all-male committee did its own study three years later.

Bartlett's meat inspection campaign and her street-cleaning demonstration attracted national attention. Speaking invitations flooded in, and cities around the country sought her advice. In response to this need she developed a detailed procedure for sanitary inspections of cities of various sizes that she followed when doing her surveys.

She accepted consulting assignments only under specific stipulations. The request had to come from a broad constituency including elected officials, social workers' groups, boards of health, and merchants' associations, not just from women's groups. The municipality had to send her a map and provide answers in advance to a questionnaire comprising eighty-two items—including, for example, information on form of government, population size, schools, method of collecting and disposing of garbage, water supply and sewage treatment systems, tax rate, and provisions for street cleaning and repair. She required in advance copies of local ordinances relating to such things as food market inspections, and she requested that copies of the city's newspapers be sent to her for two weeks prior to her visit. Her survey had to be well publicized in newspapers before her arrival, and for this she was free to (and often did) prepare news releases herself. Finally, the site of her public address at the conclusion of her survey had to be the city's largest public auditorium—not a small church or courthouse.

Bartlett's ecclesiastical training and experience had made her a compelling and persuasive speaker. Before becoming a Unitarian minister (which her father had earlier disapproved), she had been a schoolteacher, and later a reporter and feature writer for the *Minneapolis Tribune*—then the only newspaperwoman in the Twin Cities. Her experience in journalism taught her the importance of newspaper coverage for her sanitary campaigns.

Bartlett's surveys usually took three to seven days, and she insisted on being accompanied by a group of local persons able to answer her questions. Her inspections included solid waste collection and disposal practices; the cities' water supplies; and conditions at open markets and grocery stores, at public schools and hospitals, and of streets and alleys. She charged $100 a day plus expenses for her services and provided a comprehensive report of her findings.

Bartlett did sanitary surveys for over sixty cities (including Chicago) in fourteen states. The Kentucky legislature, in the year following her examination of a dozen cities in the state, produced more health legislation than the total number of bills on record up to that time. Kentucky also provided funds for a new bacteriological laboratory and for the training of local health officials. Municipal improvement leagues intended to implement her recommendations sprang up in at least twenty other cities that Bartlett had surveyed around the country.

MARY ELIZA MCDOWELL (1854–1936)
Chicago's Garbage Lady

The stench from the stockyards, from a nearby city dump, and from "Bubbly Creek," a branch of the Chicago River that had become a stagnant open sewer, boiled up and through her windows as Mary McDowell—social worker, reformer, and organizer—settled into a tiny second-floor tenement apartment in Packingtown on Chicago's South Side in 1894. Stockyard laborers lived in the drab and treeless Packingtown, also known as "Back of the Yards." Their shabby dwellings lacked even sewer connections. Slaughterhouses, the scum-covered Bubbly Creek, and garbage dumps containing refuse hauled in by carts from many other neighborhoods surrounded Packingtown. The community suffered an undue amount of illness and a high death rate.

Fifty-year-old Mary McDowell, who eventually became known as Chicago's "garbage lady," had agreed to leave her home in Evanston, Illinois, and (on the recommendation of Jane Addams of Hull House) become director of the Packingtown settlement house project, a social experiment established by the University of Chicago.

Undaunted by the appalling sanitary conditions around her, McDowell tackled her assignment by first getting acquainted with the residents and earning their acceptance and respect. This she did over a period of years by organizing clubs for adolescents and a day-care nursery for preschool children, as well as setting up a playground and a gym. The settlement headquarters soon moved from McDowell's small apartment to a larger space over a feedstore on her street—Gross Avenue. She offered classes in music, art, and English, and established a summer camp for workers.

During an acrimonious stockyard strike in 1904, McDowell publicly supported the strikers—something no other prominent individual dared to do. On several occasions during the strike she forestalled violence through mediation. She helped organize the National Women's Trade Union League and served as president of the Chicago branch for several years. In that role, McDowell persuaded the federal government in 1907 to investigate the conditions of women and children in industry. She campaigned for wages and hours legislation in Illinois and several other states. She helped secure the establishment of the Women's Bureau of the U.S. Labor Department. Over the years McDowell, who never married, also became an active member of a number of other groups, including the National Association for the Advancement of Colored People, the League of Women Voters, the Immigrants Protective League, and the Urban League.

Finally, in 1909, almost fifteen years after she began her work in Packingtown, McDowell persuaded a few women to go with her to city hall to see Chicago's commissioner of health about the garbage dumps in their community. He passed them along to the commissioner of public works, who told them that, even though he sympathized with their problem, his minuscule budget for solid waste management tied his hands. He did suggest, however, that they launch a campaign to arouse public opinion about the deplorable conditions in Packingtown, thereby forcing the city council's finance committee to increase his commission's funding.

So McDowell unleashed an intense publicity effort. She spoke to any group who would have her. Following her talk at the Hyde Park

Presbyterian Church, a judge promised to issue an injunction prohibiting garbage dumping on Lincoln Street, the primary dump site in Packingtown. And the city health commissioner, who had earlier brushed her off, sought her out. Although he claimed that there was nowhere else to cart the garbage, he did promise to disinfect the dump before winter and to halt the sending of organic refuse there the following year. But nothing changed.

Meanwhile, McDowell urged women's clubs in Chicago to organize waste committees, and she herself became head of the City Waste Committee of the Chicago Women's City Club. This club underwrote a trip to Europe, during which McDowell learned of the sanitary practices of a number of German cities. Frankfurt's incineration plants, operating in the midst of beautifully landscaped gardens, particularly impressed her. An official told her during a tour that the plants took ugly stuff and turned it into something beautiful and useful. Heat from the incinerator generated part of Frankfurt's electricity. When McDowell came home, she recounted her findings to Chicago's clubs and social groups, including the Chicago Federation of Labor.

Her publicity campaign also exposed the corruption of one of Packingtown's aldermen, who took money in exchange for allowing garbage dumping in clay pits he owned. Following the negative publicity, Chicago, at a large annual cost, began shipping its garbage to a private reduction company. Although McDowell agreed that this was better than dumping waste at Packingtown, she envisioned a less expensive solution. She recommended that a commission be established to prepare a citywide plan for waste disposal. But nothing happened until July 1913, when the women of Illinois obtained the right to vote. Within a week, McDowell, as leader of the City Waste Committee of the Chicago Women's City Club, and several of her colleagues again approached the city council about creating a waste disposal plan. This time the council listened, and a couple of weeks later appointed a study commission. McDowell and ten others served on the panel, which engaged two experienced engineers. In its final report, this Chicago City Waste Commission recommended that the city itself own and operate the waste management facilities and engage a qualified technical staff to run the system.

In 1914 the city council adopted the report and recommendations.

Soon, as a first step, the city built a small waste reduction plant in Pack-ingtown. With the dumps gone, health got better and the death rate fell. "Garbage Lady" McDowell acquired a second byname—"Angel of the Stockyards." Although the city modified the overall plan before construc-tion of the incinerators began, Chicago in 1914 finally took an official stand against refuse dumping practices deleterious to public health and started collecting and disposing of wastes in a scientific, sanitary manner.

In spite of McDowell's ongoing complaints, there was no effort to clean up Bubbly Creek until 1906, when Upton Sinclair shamed the city council into action with his novel about Chicago's meat-packing industry, *The Jungle*. The city then built a proper sewer system and filled in the creek, and people thereafter referred to McDowell as the "Duchess of Bubbly Creek." In 1929, after thirty-five years, she resigned as director of the University of Chicago settlement house in Packingtown. When she died in 1936, the residents of Packingtown gave Gross Avenue a new name, McDowell Avenue.

✷⌀ Women acted as aggressive advocates for waste management reform in the late nineteenth century. In many cities, women's organizations in-vestigated collection and disposal methods and worked successfully to pass local ordinances governing them. Women of the Louisville (Ken-tucky) Civic Association produced and distributed pamphlets detailing the garbage problem to four thousand citizens. They also made a film viewed by thousands of residents, *The Invisible Peril*. It portrayed how an old hat discarded in an open dump could spread disease.

Nonetheless, waste disposal remains a severe global problem. Nu-clear wastes, the introduction of thousands of toxic chemicals into wide-spread use in the years since World War II, and population growth have compounded the severity and extent of the disposal problem many times over.

Recycling is an important step being undertaken in many communi-ties. But changes in our throwaway mentality, overconsumptive lifestyles, mass production processes, and overuse of packaging are needed to help eliminate more waste at its source. Think of the New York City trash barge that cruised six months in search of a place to dump over three thousand tons of commercial garbage. And Greenpeace disclosed in 1993 that Baltimore officials had begun negotiations with China to use Tibet as a massive dump site for their city's solid waste. Tibetans are helpless to stop Chinese authorities from ruining their country's ecology in this way, since China has kept Tibet under armed occupation for nearly five decades.[2]

Whistle-Blowers under Fire

RACHEL CARSON (1907–1964)
"That Hysterical Woman"

Early in 1962, while *Silent Spring* was appearing as a series in the *New Yorker*, the agricultural chemical industry marshaled its vast power and wealth to try to block its publication as a book. Failing that, one chemical company, Velsicol Corporation, threatened to sue Rachel Carson because of what she had said about their product, chlordane. However, her scrupulous research held, and the threat evaporated. Another corporation, Monsanto, ridiculed the book with a parody called "The Desolate Year," portraying what a horrible world we would have without pesticides. The trade journal *Chemical and Engineering News* disparaged the book,[1] as did *Science* magazine, the periodical of the American Association for the Advancement of Science. Next, the industry, treating the book as a public relations problem, invested a quarter of a million dollars in a massive propaganda campaign to discredit Carson and to repair its image. The worst sting for the unforgiving scientists was that she told all she had learned in language the public could grasp at a time of increased popular interest in scientific issues.[2]

Time and *Newsweek* trashed *Silent Spring*. The former referred to Rachel Carson as a fearmonger using "emotion fanning words," and accused her of being "hysterically overemphatic in her emotional and inaccurate outburst." *Time* took issue, especially, with her statement that pesticides in the water anywhere endanger the purity of water everywhere, calling the idea nonsense.[3] But we now know that contamination extends even to the polar icecaps. (Ironically, eight years later *Time* designated the environment as the "issue of the year.") *Reader's Digest* canceled a contract with Carson to publish a twenty-thousand-word condensation of her book, producing instead a short version of the negative *Time* magazine article. Some of her harshest critics, though, admitted that they had never read the book.[4]

Rachel Carson was not a reformer, but a gifted writer. Her poetic and compelling books about nature raised awareness worldwide. Her first book, *Under the Sea Wind*, although it received critical acclaim, failed to sell because it came out late in 1941, only a week before Pearl Harbor. Reissued ten years later, it became a best-seller.

For as long as she could remember, Carson had aspired to be a writer. At the age of ten she submitted a story to *St. Nicholas* magazine that won its Silver Badge award. Captivated by a required course in biology in college, she switched her major from English to science, thinking that doing so probably meant forsaking her hopes for a literary career. But her science training and her passion for the natural world provided vast subject matter for her writing.

In 1928 Carson graduated magna cum laude from Pennsylvania College for Women (now Chatham College). In spite of the pervasive prejudice against women in science, she managed to obtain a scholarship at Johns Hopkins, where she studied genetics and received a master's degree in zoology. Carson dreamed of writing full-time for a living, but family obligations (supporting her widowed mother and two orphaned nieces) demanded a steady income. One of the first two women professionals ever hired by the U.S. Fish and Wildlife Service, Carson was a biologist and editor at that agency for seventeen years. She established a new literary standard for government publications with a set of booklets that she wrote or edited on national wildlife refuges. She also made time to do some creative writing, mostly during evenings and weekends. *The Sea around Us*, first appearing as a series in the *New Yorker*, came out in book form in 1951. Acclaimed as a literary classic, it was a best-seller for over eighteen months and won the John Burroughs Medal and a National Book Award. Ultimately, *The Sea around Us* was published in thirty-two languages.

The success of this book permitted her to leave her agency job and write full-time. She built a cottage on the coast of Maine and there wrote *The Edge of the Sea*, published in 1955. These two books capture the enchantment of a beautiful and magical, often microscopic world unseen by most humans.

During her years at the U.S. Fish and Wildlife Service, Carson had become increasingly concerned about the hazards of DDT. As early as 1945, she had unsuccessfully queried *Reader's Digest* about publishing an article on the effects of this insecticide. But it was not until 1957, following a stream of reports of bird kills from mass aerial spraying, including spraying of residential neighborhoods, that she focused all her attention on this subject.

Carson abruptly abandoned plans for other writing projects to devote all her time to researching and writing *Silent Spring*. From the outset she knew she needed to persuade the professional scientist, often afraid to stick his neck out. This meant investigating practical alternatives to saturating the environment with pervasive doses of chemical pesticides. Biological controls especially interested her. She wrote to a friend:

> I'm convinced there is a psychological angle in all this, that people, especially professional men, are uncomfortable about coming out against something, especially if they haven't absolute proof the "something" is wrong, but only a good suspicion. So they will go along with a program about which they privately have acute misgivings. So I think it is most important to build up the positive alternatives.[5]

Of all insects, 90 percent are not only beneficial but essential as pollinators and predators in our ecosystems. Furthermore, "Of the destructive variety, more than 800 species have become resistant to one or more pesticides."[6] This keeps the extermination industry constantly bustling to invent new pesticides.

PHOTO: REX GARY SCHMIDT, COURTESY OF U.S. FISH AND WILDLIFE SERVICE

RACHEL CARSON AND BOB HINES LOOKING FOR SNAPPING SHRIMP IN SPONGE ALONG MISSOURI AND OHIO KEY, FLORIDA.

Nowhere in *Silent Spring* did Carson propose that all synthetic chemical pesticides be eliminated—only that the long-lasting chlorinated hydrocarbon pesticides be restricted, and that care be exercised in the use of the shorter-lived chemical poisons. Other scientists had already warned about the hazards of pesticides like DDT. As early as 1945 Edwin Way Teale, a writer and former president of the New York Entomological Society, who later became Carson's close friend, observed that "[a] spray as indiscriminate as DDT can upset the economy of nature as much as a revolution upsets social economy."[7] And two of Carson's colleagues at the Fish and Wildlife Service had written scientific papers about the deleterious long-term consequences of DDT.

Today, at the end of the twentieth century, scientific evidence has accumulated confirming that at least fifty-one synthetic chemicals, especially those made from chlorinated hydrocarbons, are playing havoc with the endocrine systems of wildlife and humans by disrupting hormonal functions during critical stages of prenatal development. The endocrine system's glandular network continuously monitors hormone levels in the body and secretes hormones into the bloodstream at precisely the levels required. Many synthetic chemicals (including PCBs) are considered "persistent," because they do not decay and become harmless under natural processes. These products will continue to present hazards to the unborn for a long time—in the case of PCBs, for centuries.[8]

Such long-term damage is what concerned Rachel Carson most. In 1959, while at work on *Silent Spring*, she sent a progress report to her editor, Paul Brooks, at Houghton Mifflin. It reads, in part:

> I shall be concerned less with acute poisoning, which occurs usually through accident or carelessness, than with the slow, cumulative and hard-to-identify long-term effects. It is chiefly when life-span experiments are conducted with animals that the real damage shows up. No one now can honestly say what the effects of lifetime exposure in man will be, because not enough time has elapsed since these chemicals came into use. But we do know that every child born today carries his load of poison even at birth, for studies prove that these chemicals pass through the placenta. And after birth, whether breast-fed or bottle fed, the child continues to accumulate poisons, for checks of mothers' milk, as well as of the dairy product, always show some content of DDT or other chlorinated hydrocarbons.[9]

Carson also worried about genetic damage from man-made chemicals. She said in *Silent Spring*,

> Some would-be architects of our future look toward a time when it will be possible to alter the human germ plasm by design. But we may easily be doing so now by inadvertence, for many chemicals, like radiation, bring about gene mutations. It is ironic to think that man might determine his own future by something so seemingly trivial as the choice of an insect spray. . . .
>
> Future generations are unlikely to condone our lack of prudent concern for the integrity of the natural world that supports all life.[10]

Carson advocated integrated pest management through sterilizers, natural predators, and diseases focused on the pest insects. She also urged moving away from monoculture—the practice of farming only a single crop over vast areas of land rather than a variety of plants. Monoculture in farming and forestry invites pest infestations and plant diseases.[11]

The chemical industry and the media were not alone in disparaging *Silent Spring*. Angry reactions exploded within government agencies at all levels. Members of the Federal Pest Control Review Board convened to attack the book and denounce Carson. One member remarked: "I thought she was a spinster. What's she so worried about genetics for?" Apparently most board members considered this highly amusing, but the person reporting the incident said he found the entire meeting disgusting.[12]

Despite the opposition of some government agencies—especially the Department of Agriculture—and the chemical industry's massive campaign against the book, *Silent Spring* skyrocketed to the top of the best-seller list and has since been translated into at least twelve languages. *CBS Reports* put together an hour-long documentary, "The Silent Spring of Rachel Carson," in which she herself appeared. It was broadcast even though three of the five corporate sponsors, including Ralston Purina and Standard Brands, canceled their support.

In his book about Rachel Carson, *The House of Life*, which appeared ten years after *Silent Spring* was published, Paul Brooks suggested in a footnote that the pesticide industry folks who asserted that Carson was not a biologist should have a "special corner reserved for them in the Library of Hell, equipped with a barnacle-covered bench and a whale-oil lamp, by whose light they would be compelled to read out loud from her master's thesis: 'The Development of the Pronephros During the Embryonic and Early Larval Life of the Catfish (*Inctalurus Punctatus*).' "[13]

For the most part Carson distanced herself from the controversy, believing the facts in *Silent Spring* spoke for themselves. She did, however, respond to some of her critics in a talk before the Women's National Press Club on 5 December 1962.

Another reviewer . . . was offended because I made the statement that it is customary for pesticide manufacturers to support research on chemicals in the universities. . . . I can scarcely believe the reviewer is unaware of it, because his own university is among those receiving such grants. . . . Such a liaison between science and industry is a growing phenomenon, seen in other areas as well. The AMA [American Medical Association], through its newspaper, has just referred physicians to a pesticide trade association for information to help them answer patients' questions about the effects of pesticides on man. I am sure physicians have a need for information on this subject. But I would like to see them referred to authoritative scientific or medical literature—not to a trade organization whose business it is to promote the sale of pesticides.[14]

And speaking before the Women's National Book Association, she remarked:

In each of my books I have tried to say that all of the life of the planet is inter-related, that each species has its own ties to others, and that all are related to the earth. This is the theme of *The Sea around Us* and the other sea books, and it is also the message of *Silent Spring*. . . . We have already gone very far in our abuse of this planet. Some awareness of this problem has been in the air but the ideas had to be crystallized, the facts had to be brought together in *one place*. If I had not written the book, I am sure the ideas would have found another outlet. But knowing the facts as I did, I could not rest until I had brought them to public attention.[15]

Less than three months after *Silent Spring* appeared, some forty bills regulating pesticide use had been introduced in various state legislatures. Congress acted to close loopholes in federal legislation requiring chemical manufacturers to prove safety, rather than the government having to prove hazard. The following spring (1963), President Kennedy's Science Advisory Committee reported on the use of pesticides. The advisors said that too little was known about long-term effects of pesticide poisoning on man and wildlife, and concluded, "the panel is convinced that we must understand more completely the properties of these chemicals and determine their long-term impact on biological systems, including man." The panel recommended, among other administrative actions, that there be an "orderly reduction with a view to elimination of the use of persistent toxic pesticides (chemicals that leave long-lasting residues)."[16]

Today it is hard to believe that the Swiss scientist who discovered DDT—closely related to nerve gas—received the 1948 Nobel Prize for his achievement. In Italy during World War II, DDT had been effectively used to combat malaria and typhus. No one knew then about its side effects or that the target insects would soon develop resistance to it. The environmental writer Frank Graham comments in his 1972 book, *Since "Silent Spring,"* that humans have made much trouble for themselves during the past fifty years by thinking that because the air, soil, and water appeared so vast we could dump pollutants on the environment in the belief that they would shortly be diluted to harmlessness. Nature has now buckled under this assault, and extensive global pollution is the result. Unseen, persistent poisons have built up in thousands of lakes and rivers, tainting whatever life they encounter on their way to the sea. Always, they seek their favorite element—the fatty tissue of living things. Even in the most minute traces, the poisons invade living creatures, building up through the food chain to frightening and dangerous levels.

In 1969 Sweden became the first country to ban DDT, when it discovered that human milk had been contaminated by the pesticide.[17] Shortly thereafter in the United States, the Environmental Protection Agency and the President's Council on Environmental Quality came into being; the Endangered Species Act became law; pollution-control laws like the Clean Air Act and the Clean Water Act appeared on the books; and the government banned the use of the most potent pesticides, such as DDT, chlordane, and dieldrin. Unfortunately, farmers now apply twice the amount of pesticides per year as they did when *Silent Spring* came out.[18] And questionable chemical compounds remain on the market under the assumption that they are safe until proven harmful. The burden of proof needs to be reversed.

The problem is global, and each nation must modify its laws to ensure adequate protection. Worldwatch Institute points to a hopeful trend, however, in its publication *Vital Signs, 1996:* interest in organic farming has increased significantly, with sales of organic farm products having more than doubled in the United States between 1990 and 1994.[19] And efforts to regulate pesticides are expanding. For example, an international ban is being sought on persistent pollutants suspected of disrupting the endocrine system.[20]

Although Rachel Carson did not live to witness all the changes *Silent Spring* wrought, there were things in which she could take comfort. The report of the President's Science Advisory Committee in 1963 had vindicated her work. She saw magazines that had vilified her recant. And a few months before her death she testified before a Senate subcommittee investigating pesticide use, offering her recommendations for action—her final appearance as a public policy activist.

Predictably, honors and awards were showered on Carson and her books. She said she felt especially honored to be elected to membership (in December 1963) in the American Academy of Arts and Letters. At the time, the group was comprised of fifty persons, and there were only three other women members. That same week she became the first woman to be awarded the Audubon Medal. She received her most prestigious award—the Presidential Medal of Freedom—posthumously in 1980 from President Jimmy Carter. His citation reads:

> Never silent herself in the face of destructive trends, Rachel Carson fed a spring of awareness across America and beyond. A biologist with a gentle, clear voice, she welcomed her audiences to her love of the sea, while with an equally clear determined voice she warned Americans of the dangers human beings themselves pose for their own environment. Always concerned, always eloquent, she created a tide of environmental consciousness that has not ebbed.[21]

As part of its centennial celebration in 1992, the Port Washington, New York, Public Library sent a query to twenty-two notable Americans, including President Carter, asking them which one or two books they believed had most profoundly affected the thoughts and actions of humankind in the past half century. Of the forty-three books the panel selected, *Silent Spring* was mentioned most often.[22]

Because the world listened to Rachel Carson, the bald eagle, the peregrine falcon, and other species of wildlife still survive. And humans paused to consider what they were doing to themselves and to their fragile life-support systems. That some of Carson's most dire scenarios did not become realities is undoubtedly her greatest legacy.[23]

Carson almost never granted interviews, but occasionally someone managed to get through to her by phone. Invariably, they would ask,

"And what do *you* eat?" And just as invariably, Carson would answer, "Chlorinated hydrocarbons, just as every one else does."[24]

THEO COLBORN (1927–)
Sounds an Alarm on Hormone Disruption

Rachel Carson's 1990s counterpart, Dr. Theo Colborn—already in her late fifties and a grandmother when she obtained a Ph.D.—embarked on a scientific sleuthing project that would stun the worldwide environmental community in 1996, shake up the chemical industry, and create a heated debate. She brought to light effects of persistent organic pollutants that Rachel Carson had not even imagined. As principal investigator, analyst, and constructor of the "endocrine disruption hypothesis,"[25] Colborn, like Carson before her, engaged in a monumental task of gathering, organizing into a data base, and synthesizing research information gleaned from the scientific literature and from scientists around the world. The work took seven years and culminated in the publication of *Our Stolen Future*.

For over half a century, she argues, humans have been engaged in chemical warfare against themselves. "We have dusted the globe with man-made chemicals that can undermine the development of the brain and behavior, and the endocrine, immune, and reproductive systems, vital systems that assure perpetuity," Colborn says. "You cannot escape from exposure in your homes, work places, the outdoors, or in your meeting rooms."[26] Restricting the use of a few of the most deadly pesticides—those made from chlorinated hydrocarbons—and phasing out the production of PCBs came too late. The world had already become permeated with these and other longlasting synthetic chemicals. These residual poisons in all our bodies are leftovers from hazardous synthetic chemicals that have been discharged into the environment since the 1940s. Although some are now restricted or banned in the United States, many of these dangerous chemicals are still being produced for use both here and abroad.[27]

"Remember, every one of you is carrying at least five hundred measurable chemicals in your body that were never in anyone's body before the 1920s," Colborn warns in one of her many talks. She continues, "There is now undeniable evidence that a female shares her body's burden of many man-made chemicals with her baby in her womb and during breast feeding, and these chemicals are capable of interfering with the chemicals the body naturally produces to tell the baby how to develop."[28]

Called "persistent organic pollutants" (POPS), Colborn says they "are circulating throughout the world on ocean and atmospheric currents. For example, a pesticide sprayed on a crop in Florida or Texas on Monday, under normal weather conditions, would be detected in the Great Lakes on Friday. . . . DDT, which is still being used in developing countries, [is] found in practically every human and animal tissue sampled around the world. . . . As the manufacturer stated on a DDT label over forty years ago, 'Its killing power endures.' "[29]

Colborn, a lapsed pharmacist turned Colorado sheep farmer, decided in her half-century year to pursue her lifelong wish to earn graduate degrees in ecology and zoology. A passionate birdwatcher since childhood, she had worked for years as a volunteer activist in the Colorado environmental movement, focusing on water issues in the West. Lacking professional credentials, though, she felt disadvantaged. "Without a degree behind you, it was easy for opponents to dismiss you as a do-gooder, a 'little old lady in tennis shoes.' "[30] Ignoring the skepticism of male faculty advisors, who did not know what to make of so old a graduate student who came to classes in cowboy boots, she earned a master's degree in ecology in Colorado and obtained a doctorate in zoology from the University of Wisconsin.

While working in the late 1980s for the Conservation Foundation on a project to determine how well the Great Lakes had recovered from decades-long pollution, she discovered an apparent paradox. Although the lakes appeared much cleaner, something was still far from right.

Even though people in the region were preoccupied with cancer concerns, actual cancer rates among humans and wildlife, Colborn discovered, were no higher in the Great Lakes area than elsewhere. Searching for an answer to why healthy-looking wildlife in the region had difficulty reproducing healthy young, Colborn learned that many of the wide array of chemicals contaminating the wildlife had something in common—they interfered with the hormone system, jamming signals, scrambling vital messages, and mimicking the hormones themselves.

Continuing her research into hormone disruptors after her Great Lakes study ended, Colborn gradually gathered extensive information on all the chemicals she most suspected. With the assistance of other researchers, Colborn developed her endocrine disruption hypothesis based on relevant wildlife data, laboratory experiments, and past human experience with the synthetic estrogen diethylstilbestrol (DES), as well as research on the mechanisms of the human endocrine system.

Three years into her work, Colborn received a grant enabling her to focus all her energies on making sense of the myriad pieces of information she had gathered from researchers in dozens of disciplines. The grant came from the W. Alton Jones Foundation, headed by Dr. J. P. Myers, a zoologist and expert birder. Her project "was the kind of work that almost never gets done by either the government or the universities because there is no support and no reward for doing it; nobody ever got tenure for analyzing and assessing other people's work. Yet how absurd to spend billions on individual scientific studies but virtually nothing to figure out what they collectively say about the state of the Earth."[31] The grant established a senior fellowship for Colborn at the nonprofit World Wildlife Fund, and Myers worked with Colborn as collaborator and a coauthor of *Our Stolen Future*.

Early in her literature search, Colborn had come across a study by two Syracuse University zoologists indicating that synthetic chemicals could disrupt hormone functions. Published in 1950 (twelve years before *Silent Spring*), this early warning paper appeared in the *Proceedings of the Society of Experimental Biology and Medicine* and was promptly forgotten.

Seeking confirmation of her convictions from other scientists, Colborn, in collaboration with J. P. Myers, invited twenty-one key scientists from the United States, Canada, and Europe in July 1991 to an interdisciplinary meeting on endocrine disruption at the Wingspread Conference Center in Racine, Wisconsin. Specialists from seventeen fields as disparate as anthropology and zoology shared their knowledge on normal hormone function and discussed their research on the effects of hormone-disrupting chemicals on wildlife and humans. At the end of their conference, the scientists released a consensus statement warning that, unless the suspect chemicals were controlled or their use phased out worldwide, irreversible damage to human embryonic development could become common. The scientists followed up by producing a highly technical volume containing their peer-reviewed papers. *Chemically Induced Alterations in Sexual and Functional Development: The Wildlife/Human Connection* sold for $75. Although a few corporate CEOs sat up and took notice, nothing else happened. The book was too technical for most persons to understand. To educate the public, influence government regulators, and bring pressure on the chemical manufacturers, Colborn realized, "[w]e needed a book that would tell the story in simple terms for the intelligent reader."[32] And at bottom she believed the public had a right to know.

When *Our Stolen Future* hit the market in the spring of 1996, the chemical industry's scientists challenged its conclusions.[33] The trade association—worried about how the media might report the book's conclusions—distributed a press kit of twenty-two pages refuting its hypothesis. Nevertheless, the industry is taking the message seriously; it has launched a frantic search for substitutes for the fifty-one identified chemicals that, even in tiny doses, have been found to be acting like hormones. And the Environmental Protection Agency has given high priority to investigating the problem.

"[H]ow do you break the habit of using cancer and mortality as the only endpoints of concern when dealing with the risks of man-made chemicals?" Colborn asks. "We know now that traditional risk assessment will not work in the case of chemicals of this nature. These chemicals would not have been in the environment and our bodies if it had worked." What we need, she asserts, is "a balanced, well-supported research agenda that takes a closer look at the invisible health effects that affect functions such as intelligence, immune function and reproduction."[34] She insists that the burden of proof should rest with the chemical manufacturers, not the government or the people.

Colborn had said in an interview for a women's magazine that she hoped further research would prove her hypothesis wrong. But instead, since the publication of *Our Stolen Future* research reports from scientists the world over have poured into Colborn's office. They have reinforced her hypothesis, especially about interference with prenatal brain development. She regrets that she and her coauthors did not write more forcefully on this aspect of the subject. She also believes chemical hormone disruptors may be responsible for the learning problems of some children.

A multinational and multidisciplinary group of nineteen physicians and scientists, including Colborn, convened a week-long work session in Sicily in November 1995 (after the manuscript of *Our Stolen Future* had gone to the printer) specifically to discuss the neural, endocrine, and behavioral effects of endocrine-disrupting chemicals. Their five-page statement of the session's findings was published in *Toxicology and Industrial Health*.[35] At a press conference in Washington, D.C. (in May 1996, after *Our Stolen Future* had been released), this group of scientists discussed the conclusions of their Sicily work session. They pointed out that there may be no safe threshold for the suspect chemicals. They said the public must be informed, and governments need to take swift action to stem the problem.

At the second annual State of the World Forum in San Francisco in October 1996, sponsored by the Gorbachev Foundation, USA, Colborn gave a major address at the opening plenary session. She called for establishing "an international entity to address the questions posed by a large number of scientists concerning the safety of persistent organic pollutants and other biologically active chemicals that behave as endocrine disruptors. . . . [I]t is imperative to establish this international entity as soon as possible. If the hypothesis holds, the research costs will seem infinitesimally small compared to what the costs might be if society does not heed the messages from the scientific community."[36] The Forum announced that it would launch its own global initiative on toxic chemicals and health.

In the same speech, Colborn said the good news is that human genes have not been damaged—just turned on or off at the wrong moments. If exposure to persistent organic pollutants can be eliminated, future generations can develop normally again. "Over the past five years scientists from around the world have directed their research toward the problem. Our learning curve has literally shot straight up. . . . More than one hundred nations have called for negotiation of an international agreement to reduce or eliminate persistent organic pollutants, some of which are endocrine disrupting chemicals."[37]

Colborn asserts that the work she has been doing on hormone disruptors could not have been carried out by a government agency or academic institution—too many restraints and particular agendas are involved. "I had no big career goals or need for money. My four children were grown. I was free to undertake this research without limitations on my work. But if the twenty-one scientists we invited to Wingspread in 1991 had not agreed with my convictions, I would have dropped the research."

She says what distresses her most about how humans interact with the planet are greed and short-term thinking. Does she have hope that humans will be able to save the earth from ecological ruin before it is too late? Colborn notes that some corporate leaders are beginning to realize that the survival of their companies depends on their changing their policies, products, and processes. In a sense, perhaps greed and concern for their bottom lines might be turned to advantage, she adds.

When *Our Stolen Future*—promoted by the publisher as "the book the chemical industry does not want you to read"—was released, it received hostile and misinformed editorial and op-ed coverage in the media, most notably in the *New York Times*, the *Washington Post*, and the *Wall Street Journal*. Several months later there was a turnaround, and positive book reviews and editorial coverage appeared (including a balanced book review in the *New York Times*). Some of the same periodicals that had attacked *Silent Spring*—*Time*, *Newsweek*, *Science*, and *Chemical and Engineering News*—published heedful and cautious articles about *Our Stolen Future*. Colborn notes that "some of the skeptical scientists speaking and writing against the book revealed that they had not read the book, which two of them admitted, and said they had no intention of ever reading it."[38]

By the end of 1996 *Our Stolen Future* had gone through seven hardback printings and had come out in paperback. It has been translated into twelve languages and has been well received abroad. A number of the more progressive European countries have—either by local action or national commitment—already restricted or banned the production of polyvinyl chloride (PVC), which is routinely made into numerous consumer products. Producing it is a main source of dioxin—one of the most harmful POPS. Alternatives to PVC are numerous and readily available.[39]

By the end of 1996, Colborn had cut back on speaking trips and interviews to work on research articles describing the effect of endocrine disruptors on prenatal brain development. Her impact on the chemical industry may exceed that of anyone in the twentieth century.

The *Utne Reader* named Colborn one of their 1995 visionaries— "People Who Could Change Your Life." And in March 1997, on the occasion of its twenty-fifth anniversary, the United Nations Environment Programme, recognizing women from around the world for their environmental contributions, included Colborn for her pioneering research.

MARY SINCLAIR (1919–)
Takes on Dow Chemical and the Nuclear Industry

While the manufacture and use of insecticides made from chlorinated hydrocarbons ran rampant in the decades following World War II, another deadly technology coming out of war research—nuclear energy—was also having its heyday. Without any studies to asses its long-term economic, environmental, political, or social impacts, the U.S. government plunged forward. It subsidized with billions of tax dollars the development of the commercial application of the technology that had eradicated Hiroshima and Nagasaki. Where might we be today if that investment had gone into developing solar and other alternative energy technologies?

In *Global Citizen*, Donella Meadows charges that the government has continually lied about nuclear power from as far back as 1945, when it claimed that dead sheep downwind of Alamagordo had nothing to do with nuclear tests conducted there. Experts said that fallout from atmospheric testing posed no danger, that weapons-grade material would forever be secured by the government's tight surveillance, that nuclear power would be too cheap to meter. We learned only belatedly about problems at Hanford with leaking drums, about servicemen being exposed to radiation at Bikini Atoll. We heard so many confusing stories about Three Mile Island that the public trust in official statements about nuclear power evaporated. The basic bureaucratic rules of operation are designed to blur information about nuclear power: Use obscure words to hide plain truths. Never confess mistakes, even though they are inevitable. Always pretend certainty, despite enormous degrees of uncertainty.[40]

Mary Sinclair, a scientist aware of such uncertainties, nevertheless had become an early advocate of nuclear energy. Working in the Library of Congress for several years in the 1950s while her husband earned his law degree, she had written abstracts of classified reports on nuclear energy research from the Atomic Energy Commission, and she had followed the literature in the years after that. She believed the problems stemmed from the newness of the technology and could be solved.

In 1967, after she settled in Midland, Michigan, Sinclair's interest in nuclear energy became more than scientific. She learned that the local electric utility company planned to construct a nuclear power plant two miles from her home, in an area of 100,000 persons. Dow Chemical, the city's primary employer, had contracted to purchase nuclear-generated power from the utility. Sinclair knew that there were still serious problems with the technology. She recalls, "This is when my concerns about the fate of the earth in a nuclear age really began to take ahold of me." Since she and her family, then including five children, lived close to the site chosen for the new plant, she decided to make sure that it was safe by discussing its potential problems in the community and resolving them before the plant went into operation.

The combined level of power and money represented by Dow and the local electric utility company intimidated the people of Midland. One in six residents worked for Dow. Sinclair herself had been a research librarian for the company, and she also did freelance technical writing for it. The utility and the chemical maker launched an intensive campaign in the community to promote the plant. Sinclair could not believe the picture they presented:

> What enraged me the most about the nuclear power plant was the fact that so many lies were being told to the public about such grave matters. It was the same line, about how safe, clean, and economical nuclear power would be. But I had been seeing in all the technical literature that a lot of the old problems with nuclear power hadn't been solved yet, and in fact, experts were identifying many more new ones that they hadn't even thought about before. The wastes of this technology were the worst in the world and they were forever. . . . I was enraged by the fact that without

resolving these grave issues, huge financial commitments were being made to promote and spread this technology without full disclosure to the public of the impacts to them and to . . . future generations.

To stimulate discussion of her concerns in the community, she wrote what she believed were mild-sounding letters to the *Midland Daily News*. Hostility exploded on all sides. Friends stopped talking to her and avoided her in shops and on the street. The newspaper published disparaging letters to the editor that intimated she was just a "know-nothing house-wife." And editorials accused her of stirring up controversy needlessly. After her first letter to the newspaper, Sinclair could not get a job in Midland.

The community had been fed the line that the nuclear plant was the only practical way to replace Dow's obsolete coal-and-oil-fired power plant—the implication being that, if the community did not go along, the company would move its headquarters elsewhere.

At this point Sinclair rebelled. Her rights as a citizen and her freedom of speech became as important to her as the safety of the power plant. Far from being the demagogue her detractors saw, Sinclair was intelligent, well educated, mild mannered, and soft-spoken. She grew up on a farm in northern Minnesota during the Great Depression, the daughter of Austro-Hungarian immigrants. She watched her father, who worked in the iron mines (where lax safety practices led to accidents and injuries), stand up for his principles and become a labor organizer.

Facing ridicule and social ostracism in Midland, Sinclair began intense study of the nuclear safety issue. Staying on top of a highly technical subject required an enormous effort. She worried she might be wrong. If so, she would willingly admit it.

> Living in a community with a lot of Ph.D.s, most of whom where pro–nuclear power, I had to be very careful of credibility. I knew I wouldn't last a week if I wasn't absolutely sure of my data. I've learned that many corporate decisions aren't made on the basis of engineering or scientific data, but on the basis of what provides the best political advantage and profit.

Although stacks of documents from the electric company and the Atomic Energy Commission, or AEC (forerunner of the Nuclear Regulatory Commission, or NRC), filled her mailbox almost daily, the questions she raised remained unanswered. She attended a seminar in New York City on nuclear energy where she heard objective, unfiltered information. She also learned about citizens' groups protesting nuclear energy forming all around the country (many led by women).

Dipping into money she had saved for graduate school, Sinclair photocopied the best papers from the New York seminar and mailed them to twenty scientists she knew at Dow. In a cover letter she emphasized that Midland, as an intellectual, scientific community, ought to be given the opportunity to discuss the issues she had brought up to ensure that the plant would be safe. One or two scientists at Dow called her, admitting they agreed, but said they could not speak out because it would jeopardize their jobs.

One day Dorothy Dow, the daughter of the founder of the company, came to see Sinclair. She had seen the packet of material sent to some of the Dow scientists and said that she, too, had concerns. To help inform the community on the issue, Dorothy Dow volunteered to pay for copying and postage if Sinclair would send the packets to the boards of directors of Dow and the utility company. But the mailing had no effect. Her best bet, Sinclair concluded, was to focus on enlightening the public.

Citizens may intervene during the process of constructing a nuclear power plant at two main points. One is at the beginning, when the utility company applies for a building permit from the NRC (at that time the AEC). The other is at the end, when the plant is finished and the utility company applies for an operating permit. In 1970 Sinclair organized the Saginaw Valley Nuclear Study Group, which later became the small entity named as intervenor in the building-permit hearings. "My husband was able to help me with the legal forms to use, but I had to know the substance of what I was litigating. It was a full-time, fourteen- to sixteen-hour-a-day, grueling job."

The hostility against Sinclair grew uglier. People jeered and hissed at her when she spoke at public meetings. The chairperson always introduced her as a "housewife," even though the pro-nuclear scientists were introduced as "doctor." Anonymous phone calls and letters threatened her. Her children were harassed at school, although she did not learn about it until years later. Unknown to her, they had made a covenant among themselves not to add to her worries by telling her.

Dow and the electric utility organized a massive pro-nuclear rally in Midland (joined by television personality Art Linkletter) to support speedy licensing. The corporation declared a holiday so employees could participate.

The local newspaper refused to print any more of Sinclair's letters, telling her she would have to buy advertising space to air her views. She collected articles over several months that illustrated the lopsided nature of the paper's coverage and wrote and distributed a pamphlet about it. She also informed the Federal Communications Commission about the one-sided reporting of the local radio and television stations. Fearing that their licenses might not be renewed, the stations scheduled time for the intervenors to make their case.

The licensing hearings for the construction permit dragged out over two years, until the end of 1972. In the main, the delay arose because serious problems kept surfacing throughout the nuclear industry. For one thing, emergency cooling systems failed tests in Idaho. Furthermore, studies revealed more serious radiation effects than had been previously judged possible at the level of allowable exposure set earlier. In spite of all this, the AEC granted a permit to the utility company in Midland, and construction began.

At this juncture, after several years of effort and enduring public abuse, many people would have given up the fight. But not Mary Sinclair. She remarks,

> The more I thought about it, the more I recognized the enormous implications of the development of nuclear power, the more clearly I saw that it wasn't just another technology. Having had five children at home so much of the time, my life and my decisions constantly revolved around them—what they were doing, what their future would be. I started seeing the nuclear issue in terms of them. The proliferation of nuclear power could profoundly affect their future; *they* were going to have to live with it, but it was our generation that was bringing it about. This realization shook my psyche.

Sinclair believes that because women are the bearers of life, they think more deeply than men about children's health. She observed during public meetings that

> [m]en would listen to environmental and social issues about nuclear power patiently at meetings, but when I brought out the fact that no home or business could be insured in case of a nuclear accident because of the Price-Anderson Act,[41] that really got to them. They seemed to get much more highly motivated to do something then.

The scorn of even the highly educated folks in Midland, always hinting that as "just a housewife" she did not know very much, motivated Sinclair to enroll in graduate school. Focusing on subjects important to her cause, she quickly earned a master's degree in environmental communication from the University of Michigan. Later, she earned a doctorate from the university's School of Natural Resources.

Sinclair's small citizens' group and their attorney decided to challenge the utility's construction license in court on the basis that the required environmental impact statement (EIS) had been faulty. At the same time, Sinclair kept a close eye on the construction site. She unearthed proof of sloppy workmanship and falsified inspection reports. In 1976 the court decided that the EIS had indeed been inadequate; it had failed, for example, to assess the effect of nuclear wastes on the environment and showed no investigation of alternatives like energy conservation. Even though the court ordered the Nuclear Regulatory Commission to re-examine its decision to grant the construction permit, the NRC instead simply conducted hearings on whether to stop construction—even as the construction continued. In the end, the U.S. Supreme Court, to which the utility had appealed, ruled that construction be allowed to continue, mainly because so much money had already been invested.

Mounting costs, emerging problems with nuclear technology, and construction delays began to erode Dow's enthusiasm for the electric utility company's nuclear plant. For less cost in time and money, the corporation could have had a state-of-the-art coal-fired plant, equipped with pollution-control devices. However, under threat of suit by the electric company if it pulled out, Dow backed down and simply renegotiated the contract.

In the meantime, Sinclair continued her close monitoring of construction-site activities. She discovered that, because the ground under the plant buildings had not been filled and compacted correctly, they had begun settling and cracking. By the time the operating-permit hearings (the last stage in the process) began, Sinclair and her citizens' group, acting as intervenors, had accumulated and documented eighteen reasons the operating permit should be denied.

The never-ending problems Sinclair exposed finally swayed public sentiment in Midland. The local paper editorialized that Sinclair had been justified from the beginning in pointing out problems. The NRC fined the utility $120,000 for major violations. In 1983, representatives of the NRC and the utility faced an overflow crowd at a public meeting convened to take questions and listen to comments from Midland residents. When Sinclair spoke this time, she heard cheers rather than jeers. Dow soon withdrew from its contract, and the electric company abandoned plans for a nuclear plant, eventually converting it to a gas-fired plant instead.

Sinclair was at first incredulous—she had fought so long and so hard, for seventeen years. When things got rough, her philosophical and spiritual perspective kept her going, allowing her to view ideas, not people, as her opponents. "I must say that initially I did not expect to halt the Midland nuclear plant—I just believed that if I brought out some of the unresolved safety issues I was aware of that we could improve the safety of these plants before they were completed and we had to live with them."

In 1996, Mary Sinclair is still actively involved in nuclear issues.

> After the Midland plant construction was shut down, I discovered that Michigan was intended to be sited for a so-called low-level radioactive waste facility for the seven states of the Midwest Compact. I knew neither the public, the media nor the politicians knew how misleading the term "low-level" was, since it contained many resins and contaminated reactor parts that were as highly toxic as the spent fuel from which these high-level radioactive wastes had leaked. I worked for four years to educate the public in various ways and to organize a statewide coalition, "Don't Waste Michigan," to be more effective politically. We finally got Michigan dropped from the compact.
>
> About five years ago, it was brought to my attention by a woman

from Wisconsin that Consumer Power Company was planning to put high-level nuclear waste, spent fuel, from the Palisades plant into concrete casks that had never been built or tested before.[42] These casks were to be placed on a concrete slab 150 yards from Lake Michigan. In addition, the Nuclear Regulatory Commission was going to institute its "generic" licensing policy with the installation of these casks. This meant they would not require any environmental impact statement, no site-specific studies, and allow no public hearings for these actions. This policy was established at Palisades, with the intention that it would be the policy for placing wastes in casks at every reactor site in the country. We made all kinds of appeals to have all the issues that we could see were not resolved considered by the NRC. Our attorney . . . joined "Don't Waste Michigan" and "Lake Michigan Federation" in making these requests of the NRC, but all to no avail. The "generic" licensing process is in place. High-level nuclear waste can now be placed in dry cask storage at every reactor site in the country. All of them are on the major freshwater supplies of the country. It has been impossible to get this story in the national media as it should be. . . . On May 28, 1996, there was a flash fire in one of these casks. . . . This fire had enough explosive force to raise a three-ton lid and tilt it on its side. This has vindicated that a thorough public analysis and a hearing for using these types of casks for high-level waste storage should have taken place. . . . NRC has halted the use of the types of casks used at Palisades and Point Beach indefinitely.[43] It has halted the use of all casks at other utilities' sites for forty-five days while they study and report to the NRC the impact of any chemical reactions possible in the casks that are to hold or are holding the nuclear waste.

Mary Sinclair, now in her late seventies, has persisted against formidable odds in her anti-nuclear mission over three decades. We have a safer world because of her and others like her.

The nuclear power industry is in its death throes. No new nuclear plants are being built today in the United States—none has been ordered since the mid-1970s.

DAI QING (1941–)
Confronts the Chinese Government

The prison I was put in was one for high-level political prisoners—a very famous prison in China. In China we have lots of political movements or factional fighting in the [Communist] Party. So we have many high-level political prisoners, and they are just put in prison—no trial, no trial at all, just put in prison. All the jailers know that maybe one day the political prisoners will be released and may become VIPs again. I cannot say that by U.S. standards of prisons the condition is very good, but I was never hungry and the jailers were never rude, but quite polite. There was only one prisoner per cell—a very small cell.

Could you write? Or read books?

In the beginning, no. Cannot read, cannot have broadcast, no radio, no newspaper, no books—only sit. Very difficult in the beginning. In China lots of people in prison are put there for some secret excuse by the authorities. You are sure you are innocent.

Could you talk with other prisoners?

No, no, no. Absolutely cannot. Every time we have a chance to get outside—everybody has a small courtyard, but no big courtyard together, but each separate—they had a system whereby a jailer would lead the first prisoner out to his or her courtyard and when he turned a corner and could not be seen, then a second prisoner would be led out. Could never see or talk to each other.

Dai Qing's crime? As a journalist working for a Beijing newspaper, the *Enlightenment Daily*, she became the first Chinese writer to publicize the views of opponents of the Three Gorges Dam. And early in 1989, at great personal risk, Dai assembled—and found a woman publisher in China to produce—a collection of essays, interviews, and statements by Chinese intellectuals, scientists, and writers opposing construction of the massive dam at the scenic Three Gorges of the Yangtze River. *Yangtze! Yangtze!* helped stimulate China's first public debate on an environmental issue. Copies reached the hands of delegates to the National People's Congress (NPC), scheduled to vote in April 1989 on whether to proceed with the dam. Information from government-controlled media was distorted and riddled with deceptions. Advantages of the dam had been greatly exaggerated, problems ignored, costs understated, opposition disregarded. Even during the NPC's deliberations, the government did not want opponents' voices to be heard—literally. When one scientist attempted to talk about some serious problems with the dam, the sound system went dead. Nevertheless, even though the final vote favored the dam, Dai Qing was encouraged: "I really feel good because the delegates read my book and know there is another opinion and other alternatives, and for the first time in history of the People's Republic of China, one-third of the delegates did not agree with this project. All the other times, 99 percent totally agreed with the Party's decisions. So that is hopeful."

Dai said that she had not entered the Three Gorges debate solely because of environmental concerns. She also saw it as a human rights issue involving freedom of the press, freedom of speech, and a more democratic decision-making process. Her book "marked what the *Far Eastern Economic Review* called 'a watershed event in post-1949 Chinese politics as it represented the first use of large-scale public lobbying by intellectuals and public figures to influence the governmental decision-making process.' "[44]

Dai Qing is no mere outsider or "counter-revolutionary." Her parents, both loyal Communists, had worked as intelligence agents in the 1940s, spying on the Japanese in occupied Beijing. Her father, arrested and executed by the invaders, became a revolutionary martyr. After her father's death, his comrade-in-arms Ye Jianying, one of ten marshals of the People's Liberation Army and a prominent Chinese political figure, adopted Dai Qing. Her mother, eight months pregnant in 1941, survived beatings, as well as electric and water torture, before escaping from the Japanese. The newborn Dai Qing bore bruise marks.

An electrical engineer by training, she had at one time worked in military intelligence for a guided missile firm. During the Cultural Revolution, and after a disillusioning stint with the Red Guards, Dai defied the authorities by marrying and becoming pregnant before the approved age. Along with many other intellectuals, she and her husband were sent to a rural area for "reform through labor." They spent three and a half years during the 1970s as peasants, raising pigs and "reclaiming" the land on a military farm (their baby daughter had been taken away to live with a working-class family).[45] This experience shattered Dai's last vestige of respect for and loyalty to the People's Liberation Army, which she came to see as only an armed political group.

After reclaiming her child and returning to Beijing, Dai, disenchanted with the government, jumped at the chance to work for the *Enlightenment Daily*, a newspaper for intellectuals then in its pre–Tiananmen Square heyday. Before 1989, she could write what she wanted. On the side, she began writing and publishing collections of short stories, as well as books about the real history of the Communist Party—something about which she had firsthand information. During her tenure as a columnist for the *Enlightenment Daily*, her concern about the Three Gorges Dam intensified. But the Tiananmen Square protest changed the government's attitude toward her.

> I became an investigative journalist. The authorities did not like me. That was one reason I was arrested. The other reason was my involvement in the Three Gorges Dam issue, an example of the suppression of freedom. In China every journalist has to have a particular field to write about. My field is culture and scholarship—academic subjects. Three Gorges dam project belonged to the science department in the Beijing headquarters. So it was considered none of my business. In 1989 no one dared to report the opinions of the opponents of the dam.

In 1950 there were only two large dams in China. By 1985, 18,820 had been built, half the total number of dams worldwide.[46] For decades the Chinese government had been contemplating the possibility of a big dam on the Yangtze, the third largest river in the world (after the Nile and the Amazon). Ignoring scientific evidence of engineering miscalculations, weak bedrock, risks to endangered species, the issue of sedimentation, and the serious problems of resettling a large population, proponents cited flood control and electricity generation as adequate justifications. However, the dam's capacity to control floods would be limited by geography. Floodwaters come into the Yangtze from seven tributary rivers, and the dam would control water from only one. Furthermore, the dam would take twenty years to complete and would not produce any electrical power for at least another ten years. The local people would be immediate victims; they would be moved farther up the hillsides, where the terrain is steep, rocky, barren, and unsuitable for farming. At a cost of more than $24 billion,[47] Three Gorges Dam would be the biggest on the planet. It would flood 100,000 acres of fertile farmland, submerge 13 towns and cities and 657 factories, displace over 1 million people, obstruct navigation, cause extensive damage to wildlife and the environment, destroy Paleolithic archaeological relics, and obliterate what has been called the most beautiful river gorge in the world. The project covers an area four times the size of California. The environmental situation in China is already grave, and the country's own ecologists have described the Three Gorges Dam as an "enormous environmental disaster."[48]

China's agricultural and industrial heartland lies in the Yangtze Valley. Seventy percent of its grain is produced there; the area is responsible for 40 percent of the country's industrial output. Rather than being the path to prosperity claimed by its advocates, hydropower causes the loss of millions of sustainable livelihoods and inevitably results in worsened poverty.[49]

In her Freeman Lecture at the Massachusetts Institute of Technology in April 1996, Dai Qing observed, "The Three Gorges project is the product of a decision made by a single party—the dictator . . . who has the notion of being an all-powerful leader and being able to easily create something that is humongous and great and alter nature, and that this is a reflection of his greatness."[50]

Opponents of the dam argue that the critical needs for flood control and electrical power could be met more quickly, safely, and inexpensively without dislocating people. The government could develop reservoirs and smaller power stations on the tributaries in the upper reaches of the Yangtze; reinforced dikes, along with dredging, could improve navigation. Furthermore, the people dislocated by the Three Gorges Dam could be victimized more than once. In August 1975 the catastrophic failure of sixty-two dams in eastern China led to between 86,000 and 230,000 deaths. A 1995 Human Rights Watch/Asia report stated that the collapses caused regional famine and epidemics that affected an additional 10 to 12 million people. The government suppressed media coverage of the catastrophe, and even the foreign news services failed to learn of it at the time.[51] The recent worldwide release of information about this disaster heightens concern over the Three Gorges project. Were this dam to collapse, the resulting flood would be forty times greater than the 1975 catastrophe.

Yangtze! Yangtze! received favorable coverage in the Chinese media when it came out in early spring 1989. But Dai Qing's timing could not have been worse. In April, seven weeks of peaceful student demonstrations for greater freedoms began in Tiananmen Square, culminating early in June 1989 in bloodshed, executions, arrests, and repression of dissent.

Dai Qing publicly denounced the massacre, quit the Communist Party, and was arrested in July 1989. She spent ten months in prison—six of them in solitary confinement. Authorities told her she would be executed. The charge was that her book "contributed to creating the atmosphere of dissidence and was of a subversive nature." Three months later the government banned her book and destroyed most copies.

Since being released from prison, Dai Qing has been refused employment, and she lives under constant surveillance, including a telephone tap. Nevertheless, she has continued to do freelance writing and to lobby in both China and other countries to halt completely or to scale down the dam. *Yangtze! Yangtze!*—still banned in China—was published in English in Canada and the United Kingdom in 1994. Worldwide opposition to the Three Gorges Dam is growing.[52] The World Bank, the U.S. Bureau of Reclamation, and the Export-Import Bank of the United States have now all withdrawn financial and technical support for the project. But in February 1997—ignoring environmental and human rights arguments against it—Japanese, German, and Swiss government export-credit agencies announced their willingness to provide loan guarantees to corporations making power-generating equipment that were bidding for contracts on the dam.

Lawrence R. Sullivan, a professor of political science at Adelphi University, points out obstacles to halting the dam in his preface to the English edition of Dai's book. He says the dam reflects the engineering gigantism that appeals to China's leaders. Moreover, the dam is identified with nationalism and ethnocentrism—China always wants to be "number one" in everything from walls and cities to dams.

Dai Qing suggests that her country could learn much from big-dam mistakes elsewhere in the world—for example, the Aswan Dam in Egypt, which is considered by environmental scientists to be the worst ecological mistake ever made by humans in a single place.[53] In Brazil, Australia, India, Nepal, Chile, Canada, and elsewhere, dams are being scaled down or canceled because of economic and environmental considerations.

Dai employs every possible argument in her lobbying against the dam, both in China and in the United States. She reminds proponents that U.S. dam builders and lending institutions, having run out of exploits at home, have been seeking new opportunities overseas, often in developing countries like China that can ill afford the economic burden of such schemes. The planet's river systems are being radically altered without anyone's understanding the consequences of the changes.[54] Dams have limited lifetimes and are only temporary solutions to flooding and energy generation, Dai notes; eventually they silt up and become useless either for storing water or for generating power.

With the emergence of the environmental movement and a better understanding of how river ecosystems function, dam building in the United States has fallen into disfavor. Besides, there are no potential sites left on U.S. rivers for big dams. Civil engineers became heroes after the industrial revolution, especially so in the 1930s and 1940s, when dam building and other public works projects created jobs. The damming binge in the United States lasted for forty years and resulted in the construction of thousands of dams. Federal engineers in the Bureau of Reclamation and the U.S. Army Corps of Engineers, enthralled by dams, strove to make each one bigger and grander than the last. And before the passage of the National Environmental Policy Act in 1969, such projects required no environmental impact statements or assessments of possible alternatives. In many cases, water conservation incentives and preventing development on floodplains would have been less costly and better solutions.

In *Cadillac Desert*, cited in Dai Qing's *Yangtze! Yangtze!*, Marc Reisner writes that what had started out in the 1930s as an emergency effort to create jobs in the United States and to help resettle victims of the Dust Bowl exploded into a monstrous decades-long program of water projects that the government lacked the courage to halt. The reckoning for this vandalizing of our natural resources and economic future has not even begun. So far, natural river systems have suffered the most.[55]

Is Dai Qing hopeful that human beings will save the planet from ecological ruin before it is too late?

"My opinion is—I am not very happy. I am not optimistic. Some people, and more and more people, have environmental sense, but most of the 1.2 billion people in China have no sense of nature and just want to get rich. So it is just that human beings are going to destroy the environment." She adds, however, that Chinese environmentalists are warning the people that if they behave badly toward nature, China will lose its world-citizen status. Dai also notes that her husband and her only child—

the daughter once taken from her—support her activism and her position on the Three Gorges Dam. So do her six siblings, except for one brother who is an official in the Chinese government. "When I go to my family and we chat with each other among ourselves, we give our opinions very frankly. But when this one brother arrives, no one wants to say anything—they don't want to have to argue with him, so they just all be quiet."

Dai has written more than ten books and dozens of articles. Most of them have been blacklisted by the government. Her newest book, a further exposé of the Three Gorges Dam project, will shock the government, she says, because it discloses everything the leadership wants to hide from the People's Congress and from ordinary citizens. Entitled *The River Dragon Has Come* and published in 1997, the book includes a history of China's more than three thousand past dam disasters, the most notorious being the August 1975 tragedy.

In recognition of her work, Dai Qing received a Neiman Fellowship from Harvard University in 1991, the International PEN Award and the *Condé Nast Traveller* Award in 1992, and the Goldman Environmental Prize in 1993.

Unfortunately, despite worldwide concerns about the wisdom of building the Three Gorges Dam, which China's own ecologists consider a monstrous environmental disaster, the Chinese government is proceeding with construction.

CHAPTER FIVE

*If it is true that women universally have some
fundamental sensitivity to the land and air and water and
energy with which they live, and if it is true that there is
some connection between the critical point that the earth is
now reaching in terms of resources and women's awareness
of that, then the tide of women's rage may well rise up in
response to those conditions.*

SALLY MILLER GEARHART
"The Future—If There Is One—Is Female"

The Tide of Women's Rage

EMMA MUST (1963–)
Fights British Road-Building Folly

After fastening one end of a heavy chain around her neck, Emma
Must, twenty-nine years old, her heart pounding, crawled beneath the
front end of a bulldozer and shackled herself to the earthmover's axle.
Must and a citizens' group she helped organize—the Twyford Down
Alert!—were engaged in a last-ditch effort to prevent destruction of Twy-
ford Down, a beautiful hillside and its surroundings outside the historic
village of Winchester—six miles north of Southampton in the south of
England.

Because of its singular natural habitats, historic significance, and archaeological remains, Twyford Down had officially been declared a protected area. For two decades local residents had been resisting the plans
of the British Department of Transport (DOT) to tear up the Down for
a highway extension, part of a London-to-Southampton roadway. In an
effort to intimidate the groups fighting the highway, the government had
threatened court action.

When construction began in 1992, a few local residents—Friends of

Twyford Down—refused to give up. With a group of young people, they engaged in a camp-out on the Down.

Emma Must had grown up near Twyford Down and as a child had flown kites on the hill. During her daily train commute past the Down to her job as a librarian in Winchester, Must became increasingly angry as she witnessed the land being torn apart. So she joined the protestors. Late in 1992, after bulldozers had already ripped up the historically treasured medieval trackways, the protestors—who called themselves the Donga Tribe after the ancient trackways—engaged in a three-day face-to-face confrontation with the construction workers to halt further excavation. When a law-enforcement team removed the Dongas, most local citizens assumed the fight had ended and that the protestors had lost. But Emma Must, determined to stop the road, helped organize a new group, the Twyford Down Alert!

By early spring 1993, when bulldozers returned to Twyford Down, the construction workers found hundreds of protestors on hand to confront them once again. At this point Must pioneered the radical tactic of chaining herself to the axle of a bulldozer—an action that became a hallmark of the British anti-roadway movement. She spent five hours chained to the machine.

For six months the Twyford Down Alert! staged more than fifty peaceful demonstrations at the construction site. But in June 1993 the court ruled in favor of the Department of Transport and issued an injunction against fifty-seven protestors. When a number of them ignored it during a demonstration involving almost a thousand persons, Emma Must and six others were charged with a "civil offense," arrested, and jailed for two weeks.

Unfortunately, efforts to save Twyford Down failed. But the defeat served only to further enrage Must and to strengthen her resolve to change the United Kingdom's road-building policy. The campaign to save the Down had gained nationwide publicity and focused attention on the DOT's massive undertakings. The department had earlier boasted that its road-building project would be the most extensive since the Roman Empire. The overall plan would demolish local communities and ruin hundreds of sites that had been officially protected as wildlife habitats, as well as for historical, archaeological, and scenic reasons. Emma

Must never considered quitting her cause. As soon as she came out of prison, she set about converting the Twyford Down Alert! into a much broader effort against road building called Road Alert! Sentiment against road building surged across the United Kingdom. In the spring of 1994 Must joined a nationwide umbrella group called Alarm UK. She accepted responsibility for creating two local coalitions, each of which comprised around forty smaller groups in southern England, including the South Coast Against Roadbuilding, or SCAR.

By late 1994 more than 250 road-fighting groups had sprung up throughout Britain. As a result, the British DOT abandoned about sixty road projects and deferred another seventy for at least a decade; this amounted to shelving over a third of the planned program. Subsequently, public pressure forced the country's highway construction budget to be slashed. Most satisfying of all, the British DOT publicly conceded in 1994 that new roads, rather than solving traffic problems, only generate more traffic. Now the United Kingdom is creating a transportation strategy not based on building more roads. Must demonstrated that people power could force a revolution in British transportation policy.

Rarely in the twentieth century has the United Kingdom reacted so positively to citizen outrage over an issue. The broad-based campaign included supporters from every social class and of every political persuasion. In recognition of her achievements, Emma Must received one of the 1995 Goldman Foundation Environmental Prizes.

JOANN TALL (1953–)
Thwarts Exploiters

JoAnn Tall, an Oglala Lakota living on the Pine Ridge Reservation in South Dakota, was twenty years old in 1973 when she participated in the American Indian Movement's (AIM) occupation of the village of Wounded Knee. Tall, angered that federal agents had dared to invade and attempt to take over the Lakota homeland, feared she might be killed at Wounded Knee. Huddled inside one of the few buildings in the settlement, Tall, along with other women and children, cringed as gunshots rang out around them.

Wounded Knee, a sacred tribal burial ground on the Pine Ridge Reservation, had been the scene of a massacre in 1890 when three hundred Lakota people—mostly unarmed women and children—were shot by the U.S. military. The incident began when cavalrymen disarmed the Indian leader Big Foot and his band, who were traveling from the Standing Rock Reservation to the Pine Ridge Reservation. After disarming the Lakota, the cavalry took them to Wounded Knee. There, while soldiers were searching the Indians for any undiscovered weapons, a small scuffle broke out. The cavalrymen opened fire and continued shooting until every Indian lay dead.

This bloodbath changed the lives of the Lakota dramatically. The repeated breaking of treaties in subsequent years led to the loss of their homelands. After 1890, in addition to taking much of the Indians' territory, the U.S. government suppressed the spiritual and cultural customs of the Lakota people.

During that century, the government took over Wounded Knee and allowed white settlers and a white trading post to move in. In 1973 two hundred members of AIM, including JoAnn Tall, staged a seventy-one-day sit-in at Wounded Knee to reclaim it as part of the Pine Ridge Reservation and to demand, among other things, a Senate investigation into the deplorable conditions facing Indians. During the siege, U.S. marshals and FBI agents brought in armored vehicles and helicopters, and they directed automatic-weapons fire against all the buildings, including churches, at Wounded Knee. Two Indians were killed. After the protestors had held out for more than two months, the White House finally agreed to schedule a meeting to discuss problems relating to the 1868 treaty signed at Fort Laramie, so long as the Indians gave up their occupation.

The terms of the 1868 treaty had been violated repeatedly, other promises had been broken, Indian resources had been exploited, and the Indians had been terrorized by hired hoodlums, some of whom AIM activists have linked to the Bureau of Indian Affairs (BIA). Statutes changing the provisions of the treaty had been passed by Congress without consulting the Indians, itself a violation of the treaty. JoAnn Tall and her companions at Wounded Knee in 1973 sought a White House meeting, congressional hearings about the status of the historic 1868 treaty, and an investigation of the myriad violations of the civil rights of the Oglala Lakota.

As it turned out, the government honored none of the terms of the 1973 cease-fire agreement negotiated at Wounded Knee. When hundreds of Oglala gathered on the Pine Ridge Reservation on the appointed day, 17 May 1973, to meet with the promised representatives from the White House, no one from Washington showed up. Instead, a White House staff person sent a cryptic note which said that the treaty-making days with the Indians had ceased in 1871 and that only Congress could change statutes passed since then.[1]

Following the 1973 Wounded Knee episode, JoAnn Tall became further enraged as a flood of anti-Indian legislation, as many as thirty bills a year, appeared in Congress, attacking treaties and Indian sovereignty. Worse was to come. Through her work with AIM, she learned that between 1972 and 1976 the Indian Health Service (IHS) had sterilized 42 percent of all Indian women of childbearing age. This meant that almost every fertile Indian woman entering an IHS hospital had undergone this procedure.[2] Some Indian women quoted in the government's own reports finally verified the practice. Efforts to publicize the facts failed. It was only when AIM threatened to sue the government for $300 million in damages on behalf of the victims that a senator from South Dakota forced the government to put a moratorium on the program.

For more than twenty years, JoAnn Tall has dedicated herself not only to Indian rights but to environmental activism. She is motivated by the Lakota reverence for nature. Their worldview regards "Grandmother Earth" as having sustained her children throughout millennia. To the Lakota, land cannot be "owned" by anyone, much less bought and sold.

Tall has eight children. She lives in the settlement of Porcupine on the Pine Ridge Reservation, site of the nation's poorest county. Materially poor all her life, and now suffering from rheumatoid arthritis, Tall nevertheless has never hesitated to challenge government, military, or corporate interests threatening Indian burial sites, reservations, human rights, and well-being.

For example, Honeywell, a weapons manufacturer, decided in 1987 to conduct nuclear missile tests in an ancient burial and ceremonial site in a beautiful Black Hills canyon sacred to the Indians. JoAnn Tall used

PHOTO: CLAUDIO VAZQUEZ

JOANN TALL.

her radio station, Indian owned and operated, to spread the word among her people of the threatened desecration of their sacred canyon. Then, at a community gathering convened by Honeywell to pacify the Lakota with a clever public relations presentation, Tall went after the PR man, asking how he dared to try to desecrate their church. (The Lakota consider the Black Hills their outdoor cathedral.) She told the Honeywell representatives in no uncertain terms that the Indians would not yield on the issue and intended to win the fight.

Whereupon Tall and 150 Lakota camped out for three months in the canyon, refusing to leave until Honeywell gave up its plan. Tall said that she had a vision during one of their ceremonies that symbolized seven future generations, and she believed the missile testing would mean the beginning of the end for her people. Among themselves, the Lakota had made a spiritual commitment, with the help of their medicine people, to protect the canyon. And they won the day.

In the absence of an environmental justice ethic, the communities of minority groups have all too often been literally dumped on. The waste industry found out that Indian reservations, with their sovereign status under tribal rule, had fewer regulations controlling toxic waste than did land under the jurisdiction of state, county, or city governments. Therefore, in the late 1980s and early 1990s, they targeted sixty Indian communities for setting up toxic waste sites. The industry, under the guise of "economic development," began approaching poverty-stricken Indian tribes with promises of vast sums of money and many jobs in exchange for permitting toxic waste dumps on their reservations. Furthermore, they selected tribes where a language barrier prevented the Indians from fully understanding the dangers of such dumps. Those chosen had no words in their languages for hazardous toxic chemicals like PCBs or dioxins. The proposed sites included a five-thousand-acre landfill and incinerator on the Pine Ridge Reservation in South Dakota.

This proposal infuriated JoAnn Tall. The Native Resources Coalition, which she had cofounded following the missile-testing controversy, joined forces in 1991 with other grassroots groups to organize a major conference in the Black Hills to make various Indian tribes aware of the environmental racism and economic blackmail being aimed at native people (who, as it is, own only 4 percent of U.S. land). During the conference, Tall helped organize the Indigenous Environmental Network, a coalition comprising more than fifty organizations. This coalition helps to educate and protect native people in both the United States and Canada.

Largely through Tall's leadership, the native councils, in spite of promises of big money, rejected proposals for toxic waste disposal sites at the reservations, which included Pine Ridge, Rosebud, and others nearby. Moreover, being one of the federal government's formerly used defense sites (FUDS), live bombs and shells used in war games during World War II still stud the Pine Ridge Reservation and dozens of others. A nationwide debate continues on how to clean them up.[3]

Tall plays down her own achievements, pointing to the many other American Indian women activists. She says Indian women have a long record of resistance, and points out that native history details women's roles as sustainers of the spirit and vitality of their culture.

Government and industry have already caused extensive destruction of native lands by contaminating rivers and other waterways, setting up unregulated dumps, building coal-burning power plants, strip-mining, uranium mining, erecting hydroelectric dams, and logging forests intensively, not to mention conducting more than eight hundred nuclear bomb tests. JoAnn Tall and other native women in every case have been out in front, leading the protest against such destruction.

Tall and her colleagues in AIM point out that although fewer than half a million Indians now live on reservations, the revenue from their lands adds more than six billion dollars to the gross national product of the United States. Applying the rules of capitalism fairly, American Indians ought to be among the wealthiest people in the world. But the federal government holds all Indian lands in trust, and it permits corporations to exploit their natural resources in exchange for only token royalties.[4]

The Goldman Environmental Foundation selected JoAnn Tall to receive one of its global environmental hero awards for 1993.

JOAN MARTIN-BROWN (1940–)
Feminist–Environmentalist

"The relationship between the concerns of women and concerns for the environment were not always apparent to me," Joan Martin-Brown remarks. "Where did women's issues end and environmental issues begin?"[5]

In the late 1970s, during her tenure at the Environmental Protection Agency (EPA) when it was headed by Doug Costle in the Carter administration, Martin-Brown found an answer to this question. As director of public affairs (or "public awareness," as she preferred to call it) at the EPA, she had occasion to talk with many women foot soldiers as she traveled around the country. She recalls being

> besieged by women demanding to know what the agency was doing about the problems in the communities relating to water supply, air quality, threats from waste dumps and incineration, and constant noise. Women from diverse backgrounds were as outraged as I was about the environmental assaults on their neighborhoods. They spoke of miscarriages, babies born with tumors, young children suffering from undetermined

illnesses, spouses with respiratory problems, unexplained rashes, head-aches, or impotency. Others spoke of soil erosion, the destruction of whole forests on public land, and the loss of income from tourism. In each case they cited environmental pollution as the probable root cause of the problem. At the same time, they expressed a sense of powerlessness to remedy the matter and frustration with the responses of their local, state and national governments. They wanted access to information and answers from governments and from corporate boardrooms.[6]

These women had often been told that "they were crazy, that they didn't know what they were talking about, that they hadn't gone to college and didn't understand science. Of course it was corporations trying to cast aspersions on the knowledge or the experiences of people who were the victims of federal or private installations or practices."

During her four years at the EPA, Martin-Brown launched a national outreach program to make environmental information available to women so that they might set up a network to help each other; she also created materials illustrating the connections between women's issues and environmental concerns. She instituted the agency's first publication, the *EPA Journal*, distributed to members of Congress, the cabinet, U.S. embassies, opinion leaders, and citizen subscribers. She also wrote the EPA's statement of purpose, which remained unchanged until 1992.

Encouraged by her women's networking experience at the agency, Martin-Brown wanted to form a global nongovernmental organization (NGO) committed to mobilizing women in environmental management, to advancing public education, and to promoting women's perspectives in the environmental policy–making process. "I thought if I could mobilize one-half of the human race—women all around the world to become environmental activists—that it could really serve as a critical counterpart to the power of male-dominated institutions and how they generally were not respecting the ecosystems' capacities and the environment." So in 1981, with two friends, Martin-Brown created the nonprofit World-WIDE Network, herself serving as volunteer chairwoman for a decade. The network helped women around the world share knowledge, become activists if need be, and help each other. WorldWIDE also identified women who could testify at state and federal legislative hearings, telling lawmakers firsthand about what was going on in communities in order to build cases for environmental protection legislation.

Concurrently, Martin-Brown was the full-time associate regional director/North America for the United Nations Environment Programme (UNEP). When she took the job in 1981, the U.S. government's financial support for UNEP stood at zero, a situation Martin-Brown found inexcusable. Taking responsibility for generating U.S. support, she established and headed an office in Washington and lobbied Congress successfully for financial backing. When she left in 1993, the U.S. financial contribution had risen to $22 million. Martin-Brown's UNEP assignment also offered opportunities to develop synergistic cooperative programs and events involving UNEP with the WorldWIDE Network, as well as with various other entities.

Another of Martin-Brown's responsibilities during her thirteen years with UNEP involved organizing major women's conferences (the first ever) in Asia, Africa, Latin America, and the Caribbean. This effort culminated in the 1991 Global Assembly on Women and the Environment; held in Miami, it was cosponsored by the Women's Environment and Development Organization (WEDO), founded by former U.S. congresswoman Bella Abzug. The assembly drew women from eighty-four countries, who exchanged success stories on energy and environmental entrepreneurship as well as on waste and water management. Women from the tropical rain forest worked with women cabinet ministers from Scandinavia, transcending cultural differences, creating new alliances, and hammering out sophisticated analyses of the causes of Earth's ecological crisis.

This series of women's conferences built the case for including key women-and-environment mandates in the Agenda 21 document produced at the 1992 Earth Summit in Rio de Janeiro. Martin-Brown served as an official U.S. delegate to the Earth Summit. Unfortunately, the provisions in the document affecting women have not meant authority at the negotiating table in the years since. So members of the women's caucus showed up at a special Earth Summit evaluation session at the United Nations in June 1997, the fifth anniversary of the Rio conference, to push their agenda again. Their goals included ensuring that agricultural policies promote local food-crop production, that barriers to women's equal access to natural resources be removed, and that 1 percent of development aid go to loans for rural women.

Martin-Brown acted as UNEP liaison not only with environmental NGOs but also with government agencies, the diplomatic community, trade associations, civic and religious leaders, and corporations.

Enraged by industry's assaults on the environment, Martin-Brown focused some of her initiatives at UNEP on educating the industrial sector and pulling industry representatives into dialogues about environmental issues. For example, she initiated the first World Industry Conference on Environmental Management in collaboration with the Business Roundtable. She also inaugurated a joint UNEP-industry emergency preparedness response program, working with a chemical industry trade association.

Martin-Brown's commitment to worldwide environmental education, and her training as a teacher, inspired her to undertake a UNEP project to have Dr. Seuss's children's book *The Lorax* translated into a number of languages. The book features an imaginary creature who protects trees. To implement the project, Martin-Brown entered into partnerships with the author, the book's publisher, and the governments of Thailand, Kenya, Egypt, Pakistan, Iran, Bahrain, Saudi Arabia, Brazil, China, and Tanzania.[7]

Outdoor experiences such as camping and hiking while growing up gave Martin-Brown a lifetime feeling of affinity and concern for the environment. At sixteen, while still in high school, she assisted a local chapter of the League of Women Voters with a major study on water pollution. Martin-Brown also attributes some of her environmental awareness to William O. Douglas, associate justice of the Supreme Court, to whom she was married from 1963 to 1966. But the marriage—her first, his third—did not work out. She felt subservient being just a Mrs. Somebody Important without an identity of her own. She said she especially hated all the trappings of the attendant social functions.

Following another four-year marriage and the birth of two children, she found herself again alone with no livelihood. Her life changed in 1972 when Margaret Mead invited her to dinner to meet Representative Frances Payne Bolton, a wealthy widow with a passionate concern for the environment. When Martin-Brown told Bolton that she wished to establish links between nongovernmental organizations, industry, and government to collaborate on solving environmental problems, the congresswoman wrote a check for $100,000 on the spot. The Bolton Institute for a Sustainable Future, with Martin-Brown as founding president, was born.

During the five years she ran the Bolton Institute, the operating budget grew to $3.5 million, the staff to eighteen persons. The institute began its work shortly after the National Environmental Policy Act (NEPA) emerged from Congress. The act required government agencies to submit detailed environmental impact statements (EIS) before undertaking projects that might in any way disrupt the environment. Under Martin-Brown's direction, the institute designed the first training program for federal and state officials on how to prepare an EIS to comply with the new law.

The institute focused on programs linking environmental concerns with key sectors, disciplines, and publics, from local to global. Martin-Brown says public understanding, or environmental literacy, is essential to achieving the political will necessary to solve environmental problems.

What distresses Martin-Brown most about the way humans abuse the planet? Unlike earlier civilizations, which ravaged natural resources out of ignorance, our own level of knowledge about the interconnectedness of all life and ecosystems, she says, leaves little justification for the abuse of the earth now going on.

❧ In 1993 Martin-Brown left her post at UNEP and completed work at Georgetown University for a master's degree in liberal studies that combined international relations, environment/natural sciences, and values and public policy. Writer, lecturer, and women's advocate, Martin-Brown is associated with the Council on Foreign Relations and the Goldman Family Foundation. Having been one of the World Bank's harshest critics while working for UNEP, in 1994 she accepted a position at the organization as advisor to the vice president for environmentally sustainable development. Among numerous other assignments, she represented the World Bank at the 1995 Climate Change Meeting in Rome.

"The World Bank," she says, "is going through a metamorphosis

... and is beginning to respond to a broad-based political consensus that classic economic development is not working, but is creating many social and ecological problems. Developing economies need to be responsive to social and ecological concerns if they are to be sustainable in the long run."

Throughout her thirty-five years of fighting for the environment, Martin-Brown has worried about the absence of coordination between women's NGOs and environmental NGOs. From her experience, she estimates that as many as 75 percent of community activists are women. "That's why I'm a little bit cranky that not until very recently did any environmental organization have a woman CEO." But, she points out, neither women by themselves, nor environmental groups alone, can muster enough political clout to attract policy makers' attention. She feels there has been too much compartmentalized, specialized thinking brought to solving environmental problems.

What gives Martin-Brown hope that humans will be able to save the planet before it is too late?

"I don't know if I have that hope," she says.

> The only hope I have is that the recent upsurge in interest in spirituality, the recent upsurge in the debate about values, the recent upsurge in people's repugnance toward tobacco advertising, toward televised advertising messages, toward basic tawdriness or lack of manners—all these things are beginning to give some hope that people are turning their backs on materialism and beginning to understand that their needs have been confused with their wants. .... It may be that the X-generation is the transitional generation and that their children will begin to look for another kind of value system other than that which is material. That could save the planet.

In her article "Women in the Ecological Mainstream," written in 1992 following the UNCED meeting in Rio, Martin-Brown observed:

> The dominant gender will fail in its vital efforts to set the world on a new course towards sustainable development unless that endeavor incorporates women as both social actors and integrators. In freezing out female perceptions from decision-making, we wall out the possibility of doubling the options for human ingenuity, the evolution of new values and

new ethics as well as social arrangements which could advance the well-being of life. . . .

Male perceptions, priorities and approaches to problem-solving need the enlightened perspective developed by women as homemakers, farmers, parents, spouses and entrepreneurs. Although male/female perceptions and priorities should be integrated, women have a hard time confronting the dominant mechanistic perspective, in which body and mind, nature and people, knowledge and values, present and future generations, science and society, technology and economics are segregated from each other. Women tend to "think" in "different" contextual arrangements in their inter-relationships with the environment. . . .

Women and the environment are the "shadow subsidies" which support all societies. Both are under-valued or perceived as free even while others profit from them. . . . The current hierarchical, separatist, competitive, predatory, social and economic arrangements which dominate in most societies are antithetical to a realization of the contributions of nature and women. . . .

The political future of women is to lead, for the right reason, in the right direction. The right reason is human survival, and the right direction is towards a better future for people on an Earth which can support life. Politics is an approach to managing power. When properly managed, power enables people to make a difference for the better in their own lives and serves human justice and survival.[8]

CHAPTER SIX

*Reform is born of need, not pity. No vital movement of the
people has worked down for good or evil; fermented,
instead, carried up the heaving, cloggy mass.*

REBECCA HARDING DAVIS
"Life in the Iron Mills"

"Those Know-Nothing Housewives"

LOIS GIBBS (1946–)
Organizes the Victims of Love Canal

The exposure of more than a thousand families at Love Canal in
western New York to toxic chemicals generated a level of blue-collar activism unmatched anywhere in the United States, awakened the world
to chemical-dumping hazards, and led to federal legislation creating the
Superfund. At the forefront of these events were working-class housewives—none of them environmental activists, previously—galvanized by
threats to the health of their children, born and unborn. Their efforts
demonstrated that with the well-being of present and future generations
at risk, ordinary people can seize control of a situation and become smart
in a hurry.

One of these Love Canal housewives, Lois Gibbs—a bricklayer's
daughter in her mid-twenties at the time—lived three blocks from the
toxic-waste dump site. She mobilized a neighborhood committee in the
spring of 1978 that evolved that summer into the crusading Love Canal
Homeowners' Association (LCHA).

Gibbs's six-year-old son, Michael, only four months after starting school in 1977, had begun suffering from asthma and epilepsy, as well as liver, urinary, and respiratory problems. High rates of cancer, birth defects, and miscarriages devastated the community. On a single block, four families had babies born with major abnormalities. Adults also complained of liver ailments and nervous system disorders.

Lois Gibbs and her neighbors had not known when they invested their life savings in the modest homes adjacent to Love Canal in Niagara Falls, New York, that the development was perched on the periphery of a toxic waste dump. During more than a decade, Hooker Chemical Corporation had dumped 21,800 tons of toxic wastes into the partially dug and abandoned canal, a half-mile-long graveyard for such deadly chemicals as benzene, toluene, lindane, dioxin, PCBs, and chloroform. Moreover, according to an internal report, Hooker Chemical had often released toxic gases, mercury, and chlorine into the air and dumped toxic pesticides into city sewers.

New York State health officials discovered the deadly chemical dioxin leaching from Love Canal and moving through groundwater. A drinking-water system containing only an ounce of dioxin can kill millions of people;[1] Hooker Chemical's three toxic-waste dumps in the Niagara Falls area contained more than two thousand pounds.

Longtime Love Canal homeowners, living farther from the dump, had noticed toxic residues seeping to the surface as early as 1958, when three children suffered burns from the chemicals. Skin rashes were common. After filling in and covering over the dump in 1953, Hooker transferred the sixteen-acre property to the Niagara Falls School Board for one dollar, writing into the deed a clause absolving the company of any liability for chemical wastes. Ignoring the advice of its attorney to have a chemical consultant investigate the site before accepting the land, the school board built an elementary school directly on top of the dump and sold the adjoining lots to a home builder.

In the mid-1970s, Lois Gibbs and her neighbors became alarmed when they noticed a choking stench and observed sludge oozing out of the ground. Concentrations of poisonous chemicals leaked into the residents' basements and surfaced in their back yards. As a result of several years of heavy snows and rains, underground waterways, into which tons of toxic wastes had been dumped, became swollen and brought the wastes to the surface. Residents discovered rusted-out metal drums popping up in their back yards and on the school playground.

"As early as 1976," Gibbs writes, "a Calspan report had stated that dangerous chemicals were leaking out of Love Canal. Calspan, a private research laboratory hired to test in the area, had made numerous recommendations, but none had been followed."[2]

Although a local reporter wrote more than a hundred stories about the situation at Love Canal in 1977 for the *Niagara Gazette*, help did not come to the community quickly. Both city and county authorities failed to act. That Hooker Chemical was a large employer and taxpayer in economically depressed Niagara Falls doubtless had something to do with it. The mayor worried about bad publicity and the loss of tourism. State and federal agencies lacked knowledge about the health effects of the chemicals; furthermore, they had neither the legal authority nor the budget to buy out homeowners. No laws existed stipulating that if someone makes a house uninhabitable the state must buy it. The state appointed a task force to deal with the problem, but, Gibbs noted, it included no physicians, scientists, or women.

Meanwhile, Gibbs—feeling that the state agencies were being less than forthcoming and having read the many articles in the paper—launched an intense campaign intended to force local, state, and federal agencies to investigate the problem and expedite action. "We pay the salaries of [state employees]. They're supposed to be working for us. But they were treating us as if we were an inconvenience or the enemy or small children they didn't have to explain anything to" (*LC*, 56).

Michael Gibbs's maladies paralleled exactly those reported in the newspaper articles. With corroborating statements from two physicians in hand, his mother appealed to the school superintendent to allow him to transfer to another school. The superintendent refused to approve the transfer, declaring that although he did not believe the area to be contaminated, if that was the case then all the children should be removed.

"I was furious," Gibbs recalls. "Like many people I can be stubborn when I get angry. I decided to go door-to-door and see if the other parents in the neighborhood felt the same way" (*LC*, 12). She composed a four-line petition demanding that the school be closed.

Her knees shook as she approached the first house to begin obtaining signatures on the petition. "I was afraid a lot of doors would be slammed in my face, that people would think I was some crazy fanatic. . . . I had never done anything like this before" (*LC*, 13, 19). When no one answered her knock at the first house, she quit and went home. But sitting at her kitchen table, she asked herself, "What if people do slam doors in your face? People may think you're crazy. But what's more important—what people think or your child's health? Either you're going to do something or you're going to have to admit you're a coward and not do it" (*LC*, 13).

The next day Gibbs set out with new resolve, beginning in her own block with people she knew. She thought it would be easier to be brave with them. Along the way she enlisted help in obtaining signatures—a friend who had had three miscarriages. Within a short time 161 residents had signed the petition. One woman did slam the door in Gibbs's face, refused to sign the petition, and denounced her for ruining property values in the community. Gibbs later learned that other neighbors gossiped about her, accusing her of being a publicity seeker and blaming her for "rocking the boat."

Going door-to-door, Gibbs listened to neighbors' accounts of their many health problems. "The more I heard, the more frightened I became. This problem involved much more than the 99th Street School. The entire community seemed to be sick!" (*LC*, 15). One retired employee of Hooker Chemical, whose face was badly disfigured by chloracne, a symptom of dioxin poisoning, told her that he wanted to file a class action suit against Hooker, but feared if he did it would jeopardize his pension.

With the help of Beverly Paigen, a biologist, geneticist, and cancer research scientist at the Roswell Park Memorial Institute in Buffalo, Gibbs and other housewives conducted a survey documenting the health problems in the community. At first the state authorities discounted the information in the survey because it had been collected by housewives instead of epidemiologists.

When Gibbs's son had a urinary problem, she asked a physician at the New York State Health Department to do a urinalysis. He refused, "because the New York State Health Department's laboratories did testing for Hooker, and there might be a conflict of interest there" (*LC*, 89).

But early in the spring of 1978, after its own inspection, the New York State Health Department announced that the dump site did pose a serious threat to residents' health. At a public hearing one woman told of her dog's nose being burned after sniffing the ground, and a father said the soles of his young daughter's feet had been burned by walking in his back yard. New York's Department of Environmental Conservation (DEC) then conducted a house-to-house survey, testing the air and soil, taking samples of residents' blood, and requesting that they complete a long questionnaire. The agency also arranged to put monitoring equipment in some of the basements.

The state's report of its findings, issued in April 1978, disclosed that exposure to the chemicals involved could harm every human physiologic system. The report documented that at Love Canal, between 1958 and 1975, five out of every twenty-four children had been born with defects—including deformed ears and teeth, deafness, cleft palates, and mental retardation, as well as abnormalities of the kidneys, heart, and pelvis. The report also substantiated the abnormally high rate of miscarriages in the community (50 percent higher for women living there). The Health Department warned residents to stay on the sidewalks, not to eat vegetables grown in their own gardens, and to stay out of their yards and basements.

On 7 August 1978, President Jimmy Carter declared Love Canal a federal disaster area, the first such designation for a situation not involving a natural calamity.

That same month, Gibbs presented her school-closing petition at a State Health Department hearing in Albany. She crammed for the hearing, reading every article she could find. Gibbs recalls, "I realized that I had to know what I was talking about. . . . I was intimidated by the meeting—me, Lois Gibbs, a housewife whose biggest decision up to then had been what color wallpaper to use in my kitchen" (*LC*, 27). At the conclusion of the hearing, the Health Department ordered the elementary school closed. It also recommended that children under two and pregnant women closest to the dump site—those living in the first two rings of homes—be temporarily evacuated.

With that announcement the last trace of Gibbs's shyness evaporated. For the benefit of the reporters at the hearing, she shouted at the health commissioner, calling him a murderer and crying out, "If the dump will hurt pregnant women and children under two, what, for God's sake, is it going to do to the rest of us? We can't eat out of our garden. We can't go in our backyard. We can't have children under two. We can't be pregnant. You're telling us it's safe for the rest of us?" (*LC*, 30).

When Gibbs arrived home from the Albany hearing, exhausted after driving all day, she found a near riot in the street, with screaming residents about to burn their mortgage and tax documents. They had heard about the Health Department's announcement on the radio and plied Gibbs with questions for which she had no answers. All she could say was that Health Department officials would be in Love Canal the following evening to respond to their concerns.

State officials faced an angry, shouting, overflow crowd in the school auditorium the following evening. "Everyone was emotional, myself included," Gibbs remembers. "Everywhere you looked someone was crying, or hysterical or near panic" (*LC*, 37). She said that never before had she seen Love Canal he-men cry. People objected to "temporary" relocation, which would split up families and leave some persons in the most hazardous zone. Before the meeting ended late in the evening, the state officials agreed to keep families together under the temporary relocation plan.

After the meeting reporters began calling Gibbs for statements. Love Canal had reached the nation's headlines. Her phone rang until 3:00 A.M., and the calls began again three hours later. Gibbs had never had to deal with the national media before. "It was all a new experience . . . exciting but also frightening. I wasn't sure what to do next or how to do it. . . . I feel embarrassed about some of the statements I gave [the press]. I have a limited education, and my vocabulary isn't that large. I didn't know quite what to write or say. When anybody asked me for a comment, I felt inadequate" (*LC*, 38, 43). Gibbs later admitted that in the beginning she had not even known what the initials EPA stood for.

In the ensuing months Gibbs would organize rallies and stage events to gain media attention. On one occasion, as a way of staying in the news, she and the other homeowners constructed a child's coffin and delivered it to the governor's office in Albany. She appeared twice on Phil Donahue's television talk show and once on NBC's *Today* show. She conducted numerous press conferences, testified before Congress, and met with heads of federal agencies.

Meanwhile, the residents of Love Canal had become so frustrated with the government's slowness to act that Gibbs feared the situation might lead to violence. At this point, she and some of her neighbors decided to set up a formal organization. In consultation with the lawyer who had been helping the residents, they formed the Love Canal Home-owners' Association (LCHA) in August 1978, with Gibbs as president. About five hundred persons signed up.

Gibbs had never conducted a meeting before; she did not know what motions were or how they were voted on; she did not know how to moderate discussions; and she had never heard of *Robert's Rules of Order*. At first without benefit of bylaws or organizational structure, the association set as its goals permanently relocating everyone who wanted to leave the development, doing something to shore up property values, getting the dump contained safely, and having the air, soil, and water tested throughout the area to see how far the contamination had spread. Gibbs and other unpaid volunteers ran the LCHA office in the school building. The group raised money through donations, dues, raffles, rummage sales, cake bakes, and speaking fees.

Not everyone liked Gibbs or agreed with her approach, and a number of competing citizens' groups sprang up in the community, each vying to speak for the residents. As dissensions arose, Gibbs noted that emergencies bring out the worst as well as the best in humans.

Nonetheless, activism was on the rise. Women who had in the past looked down on demonstrators and picketers—doing radical things and being arrested—found themselves doing those same things. This meant that laundry, vacuuming, dinner, and husbands often had to wait. But the activists believed that if the waste site presented a danger to pregnant women and infants, then all residents were at risk.

The state eventually identified over two hundred chemicals in the Love Canal soup, including carcinogens and chemicals associated with visual defects, deafness, respiratory and cardiac arrest, convulsions, acute leukemia, skin irritations, central nervous system depressions, and renal damage.

Following the release of the Health Department's report, Governor

Hugh Carey—who was running for re-election while being badgered by Lois Gibbs and the Love Canal Homeowners' Association—announced in August 1978 (without yet being sure where the money would come from) that the state would purchase the homes of the 237 families living closest to the dump site, enabling them to relocate.[3]

The 710 families left behind, Gibbs's included, endured the gagging stench, the stinging eyes, and the irritated skin as best they could. They had no way of knowing what further dangers they faced during the ensuing multi-million-dollar containment, cleanup, and remedial construction work at the dump. Mail carriers wore gas masks while they were in the neighborhood.

Residents demonstrated to insist that the containment work cease until everyone could be evacuated. Arrested while picketing to halt the state's containment operation, Gibbs found herself in jail for several hours. The cell had an asbestos ceiling. "All I could think of was, When is this ever going to end? I come from an environmental disaster and I'm put in an environmental disaster" (*LC*, 86).

The New York legislature appropriated money to move the residents temporarily to motels outside the work area. When the appropriated funds ran out after nine weeks, residents had no choice but to return home.

Reporters frequently asked Gibbs why the people did not just move away. Most residents could not afford to relocate on their own. Poor to begin with, they had put all their savings into their homes, now worthless. Real estate agents declined even to list them. Anger, fear, frustration, and a sense of helplessness permeated the community.

Under pressure from Gibbs and the LCHA, and on the advice of scientists, the New York authorities announced in November 1979, even before matching federal funds had been appropriated, that the state would buy the houses of all residents who wished to relocate. The state created a revitalization agency to purchase the homes, rehabilitate the area, and try to resell the houses. The mechanics of accomplishing the purchases required several more months.[4]

Then an EPA report, released in late May 1980, disclosed, among many other health problems, that some residents had damaged chromosomes.[5] This information triggered panic and a near riot in the community. Gibbs and the other distraught residents issued an ultimatum to the White House and held two EPA officials hostage for five hours in the office of the Homeowners' Association to force the government's hand. Gibbs told authorities the LCHA staff was "protecting" the EPA people from an angry mob of 250 persons demonstrating outside. A police escort had to take the officials to their car.

President Jimmy Carter had visited Niagara Falls and sympathized with the Love Canal residents and their plight. On 21 May 1980, two days after the EPA hostages were released, President Carter for the second time declared the community a federal disaster area; he also signed an emergency order providing funds for temporary relocation for all the remaining residents. The Federal Emergency Management Agency (FEMA) released $5 million to put the families up in motels—the first time in history FEMA funds had been spent on a human-made disaster.

In the meantime, enabling legislation for permanent relocation passed the New York legislature and the U.S. Congress. President Carter came back to Niagara Falls to sign the bill passed by Congress, which authorized the evacuation of all families wishing to leave permanently and provided matching funds allowing the state to purchase their homes. The Love Canal residents, New York State, and the U.S. Justice Department all filed lawsuits against Hooker Chemical.[6]

Some months passed before the federal matching funds became available and all the purchases could be arranged. In the meantime, the authorities cautioned residents that, if they had to go back to their homes for any reason, they should not spend more than an hour a week in their basements, not let children put anything in their mouths that had fallen on the ground, and not eat anything from their gardens.

Lois Gibbs demonstrated that amid nightmarish turmoil and rage, an individual can become more than she had been. She learned how to deal with bureaucrats, lawyers, professors, scientists, and nationally known reporters; she learned how to face down mayors, governors, senators, and the president of the United States; and she learned about toxicology, law, politics, and engineering. Through the efforts of Gibbs and her colleagues, the nation's aroused consciousness of hazardous wastes prompted the national Superfund legislation and the identification and cleanup of dump sites around the country.[7] In 1982, CBS broadcast a two-hour docudrama chronicling her experience.

Lois Gibbs's marriage, like those of a number of her neighbors, did not survive the stress of the Love Canal ordeal. She says she "outgrew" her husband. She moved to Arlington, Virginia, in 1981 with her two children (both soon restored to health). In response to three thousand letters she received requesting help in solving toxic-waste problems in communities all around the country, Gibbs founded the Center for Health, Environment and Justice. She works with eight thousand grass-roots groups, provides information and technical assistance, and conducts workshops, teaching communities near toxic dumps how to mount successful protests and engage the media. In collaboration with the Natural Resources Defense Council, the Sierra Club, and several other environmental organizations, Gibbs attempted unsuccessfully in 1990 to block the resale of the Love Canal houses.

In recognition of her leadership, the Goldman Environmental Foundation selected Gibbs to receive one of its six inaugural environmental awards in 1990.

CATHY HINDS (1952–)
Battles Poisons in Maine

Three states east of Love Canal, another young blue-collar house-wife was confronting pollution troubles of her own. In 1975 Cathy Hinds and her family, seeking a rural environment with clean air and water, moved from Portland to East Gray, Maine, a small community to its north.

The foul smell and taste of the well water in East Gray assaulted Hinds's senses immediately. But she rationalized that well water in the country typically contained iron and other minerals. Complaints from neighbors about foul water led Hinds to believe that all the community's wells drew from the same underground source. Before long, though, her entire house reeked of the water's odor. And her two small daughters complained of their bath water being too hot, even when it was lukewarm.

The girls broke out in skin rashes. Her husband began suffering from asthma.

Two years after moving to East Gray, Hinds had a miscarriage. (In an early warning signal, her cat had stillborn or deformed kittens.) And throughout 1977 her eldest daughter experienced dizzy spells, sometimes causing her to fall off her chair at the dinner table. Repeated trips to emergency clinics and doctors failed to reveal any cause for the problems. Hinds says that when she asked a doctor if the family's growing health problems might be related to their well water, he shrugged off the idea and prescribed tranquilizers.

Hinds's growing concern impelled her to organize neighborhood meetings to talk about the problems. She invited the health officer from the larger town of Gray to participate in the sessions. He sent water samples to several testing laboratories. Because the technology for detecting toxic chemicals had not yet been developed everywhere, several months elapsed before a testing laboratory in Massachusetts identified three contaminants in the water. Two of them, trichloroethylene and trichloroethane, were toxic but odorless. The third contaminant, dimethyl sulfide, although not toxic, luckily produced the odor that signaled trouble. Subsequent water tests detected dozens of contaminants.

As was the case in New York, the health department in Maine lacked any understanding in the 1970s of the health effects of toxic chemicals. In fact, a Maine state laboratory, after testing the East Gray water, approved it for drinking. Meanwhile, the local health officer had been conducting research on his own and feeding his findings to Hinds. First, he recommended that residents not drink the water. Several months later Hinds learned that cooking with the water could also be dangerous. Finally, baths, laundry, and household cleaning, including flushing toilets, became inadvisable—tests showed that the water gave off toxic fumes.

Dozens of wells in the community had to be capped in January 1978. The authorities traced the source of the contamination to the McKin Company, a firm down the road from Hinds that handled industrial wastes for three hundred firms. The contaminants had seeped into the underground water supply. Though state officials ordered the company to close, the chemical leftovers in its waste dump remained a presence.

The everyday problems associated with having no running water threw Hinds's life and that of her neighbors into chaos. Beginning during a Maine winter, they had to haul water in jugs for all uses from a public spigot in the town water district. When tests determined that even it contained traces of trichloroethylene, Hinds hauled water from a source even farther away. She adopted multi-use, gray-water techniques: using dishwater for mopping floors and flushing toilets, for example. One day, she recalls, she had heated the last bit of water in the bottom of her last water container to wash dishes. Pouring it into her sink, she watched the water drain away because she had not seated the stopper tightly enough. Frustrated and angry, she collapsed on the kitchen floor and cried.

Meanwhile, Hinds's family's health problems and those in the community worsened—headaches, dizziness, asthma, rashes, respiratory troubles, and a rate of miscarriage 7.9 times the average. Puzzled doctors did not know what to do about these maladies.

Heretofore a reserved person, Hinds's growing exasperation changed her into an outspoken advocate for citizens' rights. She says, "My father . . . instilled in me the importance of standing up for what I feel is right—to stand up and take action. This is America. It is not supposed to be this way."

Following the discovery of the contaminated water, residents had been participating in all the town meetings, where officials squabbled among themselves. Hinds remembers having the impression that town officials resented the residents as though they had done something wrong, as though they were causing great inconvenience to the politicians. Hinds took a jar of her contaminated water to one meeting and, standing with a neighbor beside her, challenged the officials to drink it if they thought nothing was wrong. The water smelled so bad Hinds knew it would never get past their noses. Unbelievably, one official suggested using an old gasoline truck to haul water to the residents' houses. Another person offered a tank truck; when an argument ensued about who would pay for the driver, Hinds jumped to her feet and said she would drive the "damn truck" herself if it meant having clean water.

Hinds and her neighbor, meanwhile, in addition to talking to newspaper reporters, had been going house-to-house distributing news flyers they had prepared and photocopied, keeping citizens up-to-date on the latest developments and on the health officer's findings about chemicals in the well water. Public sympathy for the predicament of East Gray's residents spread to the broader community of Gray. Consequently, a public vote extended its water lines out to East Gray, even though everyone's taxes had to increase, including those of persons in unaffected areas. Six months after the wells had been capped, the community had clean running water.

Hinds became pregnant again and delivered a premature boy weighing slightly over one pound. He died on Christmas Day 1978, two days after birth. With liver and respiratory problems as well as kidney failure, the infant did not have a chance, Hinds says. "That painful loss really triggered my anger and lit a fire under me." She learned later that their water contained benzene, which could have caused internal bleeding leading to her reproductive tragedy.

About this same time the country became aware of the Love Canal catastrophe. Hinds watched *The Killing Ground,* a 1979 television documentary on the national toxic waste problem and on Love Canal. Armed with names of individuals and agencies mentioned on the show, Hinds made so many phone calls in the ensuing weeks that her monthly phone bill reached six hundred dollars. She found a valuable ally and source of information in Lois Gibbs, the Love Canal organizer.

Following up on tips from Gibbs, Cathy Hinds and her neighbor formed the Environmental Public Interest Coalition (EPIC). Although the group had only twenty members—including some people from other parts of Maine—Hinds realized that state environmental officers had the impression, because of its effectiveness, that the organization was much larger. It worked to the group's advantage to let them believe it.

Soon Hinds received invitations to testify about her experience before state legislative committees that were formulating environmental bills. She testified in favor of a bill requiring monitoring of toxic chemicals from the beginning of manufacturing all the way through to waste disposal. But the public committee hearing took place so soon after her infant son's death that during her testimony she broke down crying. She remembers that, although she wanted to get up and run out of the room, somehow she managed to tell her story. Later someone told her that no bill had ever been pushed through committee so quickly.

Earlier, the Gray health officer had requested that Maine conduct a health study on East Gray residents to determine possible long-term problems. The state, however, refused to allocate the necessary money. Hinds called on Beverly Paigen of Buffalo, New York, the scientist who had assisted Lois Gibbs with a health survey at Love Canal. Together with a few residents, Hinds and Paigen inspected the McKin toxic-waste dump site.

Although the operation had been shut down, nothing had been done to remove or contain the toxic contaminants. Paigen and Hinds found children playing on the site among tanks and barrels containing liquids. One child was trying to pry up the lid of an underground tank.

Shocked and enraged, Hinds and her colleagues called television and newspaper reporters to a press conference on her front porch. Hinds had never conducted a press conference and had to ask the reporters what to do. They suggested she just tell them what she wanted them to know. She held up a letter she had received in August 1979 from the Maine Department of Environmental Protection affirming that all the above-ground contaminants had been hauled away from the McKin site. Then Hinds told reporters what she and Paigen had discovered.

The next day a state official showed up. Accompanied by reporters, Hinds took the official to the dump, demanding that exposed liquids be removed and the site be fenced off to protect children. She further insisted that the work begin within two weeks. When the official replied that the job would have to wait two months, Hinds said his response was not acceptable—if the state did not begin the cleanup in two weeks, the members of EPIC would do it themselves. The official blanched at the image of residents doing dangerous work with reporters looking on.

Hinds then scrambled to recruit people willing to go onto the site if need be, lined up gas masks, and obtained permission from McKin's lawyer to enter the property. Hours before the deadline, the state official called to plead for an extension. Hinds told him no, that members of the community—all now furious—would be there the next morning with reporters on hand if the state's crew did not show up. The crew showed. Hinds learned firsthand the value of the media in forcing government action.

She points out that

[p]eople in power [in government and industry] have this thing about saying women are emotional or hysterical housewives. They say those things for a reason, because women in touch with their feelings is a very good thing. It is effective, and they don't like it. They come up with such terms to shut us up and make us feel no good. . . . A staff person of the Maine Department of Environmental Protection told me that people in the department didn't like me because I used emotional issues when I went to the press. It is an emotional issue.

Dr. Paigen continued to pressure Maine health officials to test air samples in East Gray and do a health survey of residents. She feared that chemicals might be seeping into the soil, finding their way into basements, and contaminating the air in homes. Paigen discovered from studying documents Hinds had given her that, before residents stopped using well water, the levels of trichloroethylene in East Gray homes had measured thirty times higher than the levels in Love Canal homes.

But after testing the air, the state reported that contamination levels fell within federal standards. Not until eighteen months after the East Gray residents' direct exposure to contaminants had ceased did the state perform a health survey—and did so then, according to officials, only because a couple of "vociferous" housewives insisted on it. By that time, although symptoms had abated, the long-term effects remained undetermined. The EPA later criticized the findings of the state's study.

By now, Hinds had made it her mission to raise public awareness about hazardous wastes. She continued to increase her own knowledge by staying in touch with individuals mentioned in the television documentary *The Killing Ground*, reading, and attending meetings of environmental groups. She spoke to college classes and civic groups around Maine. But her family life suffered. She and her husband reacted differently to the loss of their infant son. He wished only to withdraw, put it behind him, and move forward with their lives. Cathy Hinds, on the other hand, turned to activism to release her anger.

She continued to feel uneasy about the possibility of contaminated air, but the state refused to do any testing. Her intuition told her that even though their water now came from the main Gray system, and even though some remedial work had been done at the toxic-waste site, contamination had not been eliminated. A new neighbor sought Hinds out to learn the history of the community's contamination problems. A breeder of Siamese cats, the neighbor reported problems with stillborn and deformed kittens. Hinds had recently given birth to another son; when only a few months old, he fell ill and had to be hospitalized. Tests revealed abnormal liver function.

At this point a desperate Hinds put their house up for sale. Even if they lived in a shack somewhere, her family had to get away from East Gray. Late in 1981, however, with the house still on the market, Hinds obtained a copy of a previously unreleased report disclosing that state tests had detected cancer-causing polychlorinated biphenyls (PCBs) at the McKin dump site. Should she publicize the report or keep it quiet? Though eager to sell the house, Hinds concluded that she must be forthright with any potential buyer. She held a press conference announcing the PCB findings. The release of the report pressured the EPA to reverse an earlier decision to bypass the McKin site on their Superfund priority list. Eventually, McKin's hazardous-waste dump would rank among the worst 8 percent of such sites in the nation.

Hinds sold her house with full disclosure to the buyer, and her family moved in the spring of 1982. Within a few months, her young son's liver function returned to normal and her husband's asthma disappeared.

Hinds's husband believed that "if he moved me away from East Gray, that that would take East Gray out of me—and it never did." She not only continued to monitor the cleanup of the McKin site but she also became a toxics activist, first on a statewide level and then nationally. In the mid-1980s, after lobbying Congress and the EPA for a strengthened Superfund on behalf of the National Campaign against Toxic Hazards, she accepted a staff job as their Northeast organizer. Threats to her personal well-being haunted this assignment. "The [New Hampshire] Attorney General's office has a thick file on intimidation events that took place while I was the Northeast organizer. More than once I was chased across the bridge over the Merrimack River by a car going eighty miles an hour. . . . I also experienced threatening phone calls."

In 1991 Hinds assumed responsibility for directing a national networking project dealing with military toxics, working with people who live near Department of Defense facilities. "[T]he military is the worst polluter in the country—much worse than any commercial industry that

I have ever heard of in the almost twenty years of my environmental activism." Her organization "networks with 150 groups around the country dealing with issues like military base closings, use of uranium weapons in the Persian Gulf, production of those weapons in [U.S.] communities, industrial radioactive contamination, worker exposure during weapons production, and contamination on military bases. All bases have hazardous-waste sites." Hinds also networks with an additional five hundred groups focusing specifically on military toxics issues. Even among the military toxics activist groups, women far outnumber men, she notes.

Reflecting on her work, Hinds cites politics, bureaucracy, public apathy, and lack of funding as her primary obstacles. She believes big national nongovernmental organizations siphon off dollars from grassroots groups that really do the work in local communities, especially in poor and minority communities. There are personal tolls as well; Hinds's never-ending activism contributed to the erosion of her first marriage, which ended in 1993.

Though sometimes pessimistic about humans being able to save the planet, Cathy Hinds hopes that we can gain "eagle vision" and see the big picture. She says that in order to resolve our present ecological predicament grassroots people have to come together and say "Enough is enough!"

MICHIKO ISHIMURI (1927–)
Documents "Cats' Dancing Disease"

The past fifty years have witnessed a transformation of the traditional subservient Japanese wife and mother. Women's pent-up energies have exploded into social movements that have shaken Japanese society. Full-time housewives have assumed leadership roles in protecting the environment and their families' well-being. Akiko Domoto, a woman member of the Japanese Diet, or parliament, declares that women are the driving force behind the country's environmental movement.[8]

Rapid reindustrialization after World War II, coupled with Japan's weak environmental laws and the attitude that production took priority over public welfare, created unprecedented health hazards. The most monstrous case so far was the methyl mercury poisoning of thousands of fishing villagers living near Minamata Bay on the inland Shiranui Sea. The victims, not the offender, had to shoulder the burden of proving harm.

A shy Minamata housewife and mother turned poet and writer, Michiko Ishimuri visited one neighborhood family after another, listening to their stories, crying with them over damaged lives and bodies, holding in her arms children deformed by congenital Minamata Disease. She made notes of all she saw and heard. Her first written account of the disease appeared in 1960 in a work entitled *Cruel Tales of Japan—Modern Period.* Her 1969 documentary book, *Paradise in the Sea of Sorrow,*[9] was an immediate best-seller that received three awards.[10] It chronicles everything she had learned as she talked with devastated families poisoned by industrial wastes discharged into the Minamata Bay and River by the Chisso Corporation.[11]

The first clear-cut case of the bewildering illness surfaced in 1953; by 1956, cases had grown exponentially. Healthy-appearing children and adults, without any warning symptoms, suddenly experienced numbness, loss of muscular coordination, shaking, impaired vision and hearing, speech disturbances, and delerium. These conditions gradually worsened, causing paralysis, difficulty in swallowing, blindness, deafness, deformity, convulsions, and, in 40 percent of the cases, death. "Chewing at the back of their minds," Ishimuri writes, was "that strange mixture of fear and resignation which had become an inseparable part of their lives, [as] the villagers patiently awaited their turn to fall ill and die" (*PSS,* 226).

Of patients assembling near a Public Health Department bus to be taken to a hospital for checkups, Ishimuri says:

> The people drew near: men and women carrying in their arms or on their backs children with dangling heads and deformed limbs, easily identified as victims of congenital Minamata Disease; adult patients walking in the discombobulated way characteristic of the disease. . . . No trace of resentment, however, could be detected in the bashful manner and warm, innocent smile of the sufferers. . . . Either they were completely blind or their visual field was severely restricted. The little inarticulate cries, the grotesque look on their upturned faces and the spasmodic movements of their stiff, spindly limbs were unmistakable expressions of their delight in the prospective bus trip. (*PSS,* 10–11)

The congenitally disabled child in one family Ishimuri knew well happened to be the same age as her own son. She wrote about holding him in her arms.

> Even before he tossed the boy into my arms, Mokutarō [the nine-year-old] and I had agreed through a quick exchange of glances that I should take grandfather's place in holding him. So the boy, who really was as light as a small wooden Buddha statue, was now lying in my arms. . . . Indeed, he was so light that I didn't dare to move my knees or shift my position lest he should rise into the air like a feather blown by the wind. His stick-like arms dangled purposeless on both sides of my hips. . . . His skin was so transparent and taut that it looked like it might burst open any minute and expose his skull and fleshless cheek-bones. (*PSS*, 201, 216, 218)

Michiko Ishimuri also visited patients in the hospital.

> It was because of a nearly obsessive curiosity and a vague sense of historic mission, as well as a blind impulse to expose the Minamata Disease Incident in all its ugliness and inhumanity, that I went to visit [patients] at the Minamata Disease Ward of the City Hospital in May of 1959. . . . Up to the time of my first visit to the hospital I'd been an inconspicuous, self-effacing housewife; impractical, inclined to spend my time in useless reveries. I had a preference for old songs and ballads, and occasionally dabbled in poetry. . . . Judging by the average life expectancy of Kyushu women, it seemed to me that I could live to be over 70. However, I contemplated without enthusiasm the remainder of my life, thinking that it would continue to unroll itself in the same drab, uneventful way as the life I was leading then. (*PSS*, 137–39)

As a result of her visits with victims' families in Minamata and with hospital patients, however, Ishimuri's life soon became a crusade.

> The condition of most patients was so wretched that those who saw them were horrified. . . . Those who survived were not better off than the dead: as time went by it became apparent that the physical and mental disabilities caused by the Disease were irreversible. . . . [Patients in the] Minamata Disease Ward were suffering gruesome tortures. They were no longer able to move; they would fall from their beds in violent convulsions; unable to speak, they would tear at the walls and at their own bodies with their

fingernails, or howl "like dogs"—a phrase used by specialists to describe the prolonged, desperate, inhuman cry they emitted, the only means of communication with the world left to them. . . . Though filled with the brilliant rays of early summer, the Minamata Disease Ward on the second floor had the stench of a cavern of rotting flesh. It was a place of physical and mental torment, whose depths could only partially be gauged by the inarticulate screams of the dying patients. (*PSS*, 81, 90–91, 134–35)

The affliction, not really a "disease," was caused by eating fish—a staple for most villagers—contaminated by methyl mercury waste. Methyl mercury attacks the central nervous system as well as damaging the liver, kidneys, pancreas, and bone marrow. It also crosses the placenta, causing prenatal damage. In Minamata, researchers found high levels of methyl mercury in mothers' milk. Ignorant of the cause, however, people at first thought the affliction might be contagious and shunned stricken families. At one point the Public Health Department ordered spraying of household contents and clothing with DDT.

In fact, discovering the true cause would be difficult. The Chisso Corporation was the principal employer in Minamata and had dominated local politics. The mayor during the disease outbreaks in the 1950s and 1960s, for instance, had worked as a chemist for Chisso and held patents on several processes he had invented for the company. As early as 1925 Chisso had begun paying Minamata fishermen to keep quiet about the chemical pollution of the bay. Chisso added deadly methyl mercury wastes to its discharge in 1932.

By the 1950s fish had been seen floating on the water of the bay. Birds fell from the sky and died. Cats staggered around as though intoxicated, convulsing, whirling in violent circles, often collapsing or falling into the bay. By the late 1950s almost no cats remained in the Minamata Bay area, and rats infested the community. The similarity in cats' and humans' symptoms led the local people to refer to their malady as "cats' dancing disease." When symptoms appeared, though, it was too late, because irreparable brain damage had already been done.

Those disabled victims who were still able to talk complained that in public people gawked at them and whispered about them. They felt like outcasts. Some neighbors resented their receiving compensation, accusing them of living like kings and not having to work.

Though unaffected families (whose incomes depended on the polluting company) ostracized victims and disparaged Michiko Ishimuri, she persisted. She wanted to alert the world to what mercury poisoning looks like and feels like. Of the bay, for instance, she observed:

> [T]he bottom of the sea was covered with a viscous, malodorous sediment. If you cast your net anywhere in the sea, this mud-like matter would invariably cling to it, weighing it down. The fishermen caught nothing except this foul-smelling sediment, which forced them to clean their nets painstakingly before going home each day. The once clear, beautiful Sea of Shiranui had turned into a fetid swamp. (*PSS*, 167)

Ishimuri, who, as the daughter of an alcoholic stonecutter, had grown up in poverty, also wanted to document the social dimension:

> The Minamata fishermen, among whom the number of Minamata Disease patients steadily increased, were in especially dire straits. Many were forced to sell their nets, fishing gear and boats, and borrow large sums of money to meet their medical and living expenses. There were many households that could not even afford to buy their daily bowl of rice. Six years had passed since the appearance of the first Minamata Disease patient. . . . However, no program had been worked out so far for dealing with the pollution of the Shiranui Sea or for arresting the progress of the Disease, just as no financial aid had as yet been extended to the barely surviving fishermen. (*PSS*, 83)

Soon, Ishimuri was no longer alone. Frustrated by the neglect of Minamata Disease as a social, political, and medical issue by society, a group of fishermen staged a protest rally in the fall of 1959 at the gates of the Chisso plant. A few weeks later the company installed a cyclator that it characterized as a wastewater treatment system. It was only a hoax to fool the public. The cyclator did not prevent methyl mercury from flowing into the bay.

In 1963, Ishimuri attempted to persuade the mayor to allow her to organize an exhibition of the shattering photographs of Minamata Disease patients taken by a talented photographer. The mayor turned her down. Later, she succeeded in persuading a teachers' union to permit her to mount the exhibition in a department store as part of Education Day.

Michiko Ishimuri and several colleagues founded a small magazine in 1964 called *Contemporary Records*. The venture incurred heavy debts and folded after only one issue. Ishimuri had hoped to use this periodical to publish a sequel to her first book. Disappointed, she wrote, "Reflecting on my actions, the little writing I had done and the meaning of my life, I concluded that I was an irresolute, confused woman, often missing the point and wasting her energies in useless undertakings. . . . For a short while, I had slipped out of my rigidly determined role as housewife and mother, and had now to slip back into it" (*PSS*, 295–96).

Then in 1965 the disease appeared in Niigata, a nearby prefecture. Ishimuri sprang back into action. "Something had to be done while there was still time. I began writing letters, appeals and petitions to those who had supported the Minamata Disease patients all along, and to those likely to support our cause in the future" (*PSS*, 297). She helped organize the Association for Countering Minamata Disease, immersing herself in it for a number of years. The group, consisting of two hundred unafflicted Minamata residents, pressed the central government to issue an unequivocal statement on the cause of the disease and worked to impress on both central and local governments the need to ensure the health care and livelihood of patients.

In an article in a local newspaper, Ishimuri pointed out,

> For the first time in the fourteen-year history of this mass poisoning, . . . Minamata residents realized that some of their fellow citizens were in a terrible plight, and that they needed help. . . . [P]atients and their relatives have been continuously subjected to prejudice and discrimination and their protests and claims for compensation systematically ignored. . . . Chisso Corporation, in connivance with the concerned municipal and prefectural authorities and the central government, has constantly refused to accept responsibility for the poisoning. It goes without saying that this is a case of collective responsibility, as the various echelons of the local and national bureaucracy are as much to blame as the polluting company. (*PSS*, 323–25)

Victims' families had early on organized their own group, the Mutual Help Society of Minamata Disease Families. Chisso had thrust onto the families in this group a "consolation contract," which Ishimuri characterized as "a classic instance of capitalist industry's trampling on the most basic human rights" (*PSS*, 326). Under this contract afflicted families received a small lump sum of money as "consolation for the plight of the sick." In her newspaper article, Ishimuri adds, "The most infamous clause of the consolation contract was the one stipulating that 'even if it is proved beyond doubt that there is a cause-and-effect relationship between the wastewater from the Chisso factory and Minamata Disease, the Mutual Help Society will voluntarily abstain from claiming additional compensation' " (*PSS*, 327, 329). Further, the "consolation" disqualified victims from receiving public assistance.

Early in 1968 Michiko Ishimuri wrote a petition in the name of both the Association for Countering Minamata Disease and the Mutual Help Society to call attention to the victims' plight. The petition pleaded for three things: the patients' rights to receive welfare aid, assistance in locating employment, and special education classes at the rehabilitation hospital. The public had again become indifferent. Publicizing the petition by inserting it in the Kumamoto newspaper, the groups also presented the petition in person to the president of the Kumamoto Prefecture Assembly,[12] to the Minamata City Assemblies, and to the National Diet in Tokyo.

What the groups did not know was that Chisso's own infirmary physician and chemist, Dr. Hajime Hosokawa, had conducted his own secret research, proving in October 1959 that organic mercury from the company's effluent caused the illness. On his deathbed eleven years later he testified that when he told management of his findings at the time—recommending the company halt the contamination—his research was terminated and his documentation was burned. Meanwhile, the company regularly paid small sums to protesting victims, buying off anyone threatening to make a fuss.

Also in 1959, a group of researchers at Kumamoto University Medical School reported that their findings pointed to methyl mercury as the cause of Minamata Disease. But Chisso paid "experts" to refute the report, refusing to allow the university researchers to take further samples from its effluent. Concurrently, the Food Poisoning Section of the Ministry of Health issued a report citing organic mercury as the toxic cause of Minamata Disease. Chisso responded by saying that their own researchers had just begun an attempt to verify this theory, and until this team could assemble corroborating data the theory had to be deemed scientifically deficient. In truth, the company already had seen and destroyed Dr. Hosokawa's verifying data.

In 1968, on the day after a memorial service had been held for those who had died of Minamata Disease, Chisso attempted to intimidate citizens by announcing that, without full support of the local population, the company could not achieve its five-year recovery plan, "which amounted to a thinly disguised threat to close down the factory and withdraw from Minamata altogether" (*PSS*, 347).

Nonetheless, that summer one of Chisso's workers' unions issued a statement of unconditional support for the victims of Minamata Disease, expressing shame for having turned their backs on the suffering patients. Almost ten years after the fishermen's first protest demonstration, the union joined with them in a public rally initiated by the Association for Countering Minamata Disease. Many such gatherings and protests had occurred in the intervening decade. Members of the association had even adopted a creative method for exerting pressure on Chisso's management: they purchased shares of stock and raised protests at shareholders' meetings.

Opposing sentiments split the community. Another workers' union, which backed the mayor, continued to defend Chisso. Businessmen and industrialists organized the Citizens' Conference for the Development of Minamata City. Fifteen hundred residents attended. Chisso supporters accused the "damned weirdos" of destroying the economy of the community by requesting compensation.

Families of patients suffering from Minamata Disease boycotted the conference, just as local persons who were not patients had boycotted the Joint Memorial Service for Deceased Victims of Minamata Disease. One resident said she could survive without money from Chisso. What she would like to see was the top company executives drinking mercury effluent and suffering the agonies her family members had endured before dying. "Needless to say, this explosive atmosphere allowed little room for optimism for the pending negotiations between Disease Victims and Chisso" (*PSS*, 359).

In the fall of 1968, the Ministry of Health and Welfare and the Scientific and Technical Agency issued the government's official position on

Minamata Disease. They unequivocally placed full responsibility on Chisso as the methyl mercury polluter. The company's president then came to Minamata to visit victims and apologize. Ishimuri used all the money she had on hand to hire a taxi and follow the president from house to house.

In the spring of 1973, the Kumamoto Prefecture District Court, as the result of a lawsuit filed by victims' families, ruled that the Chisso Corporation must compensate the afflicted and their families. By 1977 Chisso had been forced to pay more than one hundred million dollars in indemnities to more than a thousand persons suffering from Minamata Disease. The government also compelled the firm to dredge and fill in parts of Minamata Bay at a cost of over forty million dollars.[13]

In 1988 the former president of Chisso and the former director of the Minamata plant were convicted of criminal negligence.

"In our modern world of progress and civilization," Ishimuri remarks, "we have long forgotten what it means to live in keeping with the laws of nature; we have become deaf and blind to the vibrant soul of all things surrounding us. How could those who measured everything in terms of charts and statistics, and who reduced human life to a series of mechanical repetitions and the human being to a mathematical quantity, understand the feelings of the Minamata fishermen?" (*PSS*, 236).

Ishimuri's 1969 book has gone through thirty printings; it has been adapted for television and has supplied lyrics for blues ballads. Playwrights have used it as a source for the stage, and high schools in Japan have added it to their literature curriculum. It has even led some readers to interrupt their own lives and travel to Minamata to help the afflicted. In the years since, Ishimuri has published several more books, along with numerous articles, poems, and pamphlets, on Minamata Disease and its significance.

Since 1972 Japanese citizen-activists have focused on preventing the construction of polluting factories. Women have played a leading role in these campaigns. Although Michiko Ishimuri no longer spearheads the Minamata movement, she continues to lecture and write in support of the victims.

In observance of the Minamata tragedy's fortieth anniversary, a citizens' group organized a two-week exposition in Tokyo in the fall of 1996.

Through exhibits of photographs and artifacts, films, dramas, lectures, and discussions—with survivors and writers and other artists participating—those attending were reminded of what had happened. The original manuscript of *Paradise in the Sea of Sorrow* was among the exhibits. And the first evening's plenary session featured a lecture by Michiko Ishimuri—"Where Are We Going (from Minamata)?"

CHAPTER SEVEN

We're swallowed up only when we are willing for it to happen.

NATHALIE TCHERNIAK SARRAUTE
The Planetarium

Combatting Death by Ecocide

At a mass rally of Russian students in Tashkent's Lenin Square late in 1989, a young woman shouldered her way to a makeshift stage and grabbed the megaphone:

"The Communist Party Central Committee has wasted all our treasures and keeps taking away the fruits of our land. A future generation is going to ask us: 'What were you doing at that time? What were you thinking about? How did you get us into this situation? Where's our wealth?' And what will we answer them? How could we look them in the eye?"[1]

Two hundred and fifty miles north of Moscow, an elderly farmer recalled with nostalgia the life of his parents before the revolution. Though poor, they had lived a decent life in a friendly community. He lamented that after collectivization, instead of everyone living as a big cooperative family, the opposite happened. People became suspicious of each other, and each just looked out for himself. In the bargain, the soil, air, and water became polluted. He referred to the result as a "big stinking ruin."

An abiding love of nature runs deep in the Russian tradition. The country is still a land of rich natural resources and resilient, talented citizens. But it is badly crippled. The peoples of the former Soviet Union were forced to stand helplessly by for almost seventy years while a military-industrial complex exploited, assaulted, and in many cases destroyed the natural resources of their country. In their book *Ecocide in the USSR*, Murray Feshbach and Alfred Friendly, Jr., assert:

> When historians finally conduct an autopsy on the Soviet Union and Soviet Communism, they may reach the verdict of death by ecocide. . . . No other great civilization so systematically and so long poisoned its land, air, water and people. . . . The growth that made the USSR a superpower has been so ill managed, so greedy in its exploitation of natural resources and so indifferent to the health of people that ecocide is inevitable.[2]

Cowed by the fear of political consequences, citizens dared not publicly protest what they witnessed. Forced to remain silent, they watched as their health, vitality, and longevity diminished to the point where half the men eligible for military conscription failed their physicals and children were born ill. In one region near an aluminum factory, the children never smile because their teeth have rotted away.

Historically, the Russian people have proven themselves to be exceedingly tough even under dreadful conditions. Inevitably, this resilient spirit began to generate secret meetings in basements, where concerned individuals organized Green Groups to plot strategies for citizen action.

TATYANA ARTYOMKINA (1946–)
Defies KGB Threats

An old Russian proverb says, "Women can do everything; men can do the rest." Tatyana Artyomkina walked into one of those covert basement meetings in the ancient, historic city of Ryazan on the Oka River in the early 1980s. She was head of the city's hydrometeorological laboratory. In that capacity, Dr. Artyomkina discovered that critical environmental and health facts were being withheld from the residents of Ryazan by the authorities.

"In all the time of working in government service, be it the hydrometeorology laboratory or the committee for nature protection, I always had at my disposal information of a secret nature. That is, there was information for 'everyone,' and information 'for official use' that was more detailed, more complete and reliable." She began passing confidential material along to the Green Group. Its members then quietly distributed brochures containing the information, and even persuaded the more democratic of Ryazan's newspapers to publish the facts. Artyomkina's environmental espionage triggered an ominous threat from the KGB. She continued, nonetheless, and in 1984 went public as a Green activist.

❧ Interested in all the sciences from an early age, Artyomkina had earned a chemistry degree at the Ryazan State Pedagogical University in 1969. After working some years for a scientific research institute, she became production engineer for a new-technology planning institute, traveling around Russia introducing the institute's creations into factories. Often, she observed, the new technology was no better than the old (sometimes it was worse), which disturbed her ethical sense. Furthermore, working in the field of chemical technology and witnessing the wasteful use of natural resources and the pollution practices of industry changed her thinking.

> I came to the deep persuasion that human civilization was simply depraved and didn't have the right to exist in such a way as I had witnessed. I began to more frequently share with my colleagues my observations and feelings, but ran up against incomprehension and complete indifference. . . . No one took me seriously. Gradually I ceased wanting to do what I was doing.
>
> On the state level [at that time], the concept of ecology was not taken into consideration. And once ecology was outside the law on the state level, it meant such problems did not have the right to exist.

Then she received an offer to work in the Ryazan laboratory of the State Committee on Hydrometeorology as director of a group establishing standards for the discharge of harmful emissions into the atmosphere. This job, she believed, "would give me the ability to realize an awakening of environmental consciousness. . . . So, in July 1981, I became a professional ecologist, although the words ecologist and ecology still didn't enter our lives."

With Mikhail Gorbachev's initiation of glasnost in 1985 came more open involvement in problem solving. Suddenly, people began to talk with each other as well as to take action. Artyomkina emerged as a powerful Green Movement advocate when she publicly (though futilely) opposed plans for constructing a tannery in the Ryazan area—the only public official to do so.

She also organized a campaign to obstruct the plans of the Ryazan city officials to lay an aboveground pipeline along the Oka River to carry away household and toxic wastes. Recruiting specialists to testify on the hazards of the pipeline, she persuaded the authorities to build an underground sewage system instead. But in 1988, before it could be constructed, a new Communist Party boss in the region attempted to resurrect the original plan. Artyomkina and her Green Movement colleagues organized a mass public demonstration against the project, after which the authorities yielded and finished the underground system.

Artyomkina's role in this episode led to her being fired from her job, and the local government closed Ryazan's only environmental monitoring laboratory, which she had headed.

The Ryazan Green Movement—headquartered for many years in Artyomkina's home—then launched an effort to establish in their community an official branch of the countrywide Committee for the Protection of Nature, planning to elect Artyomkina as leader. Their strategy sessions had led to helping elect local politicians committed to the environment, including the mayor of Ryazan. With the backing of these officeholders and the members of the Green Movement, Artyomkina defeated a number of other strong candidates and in 1990 became chairwoman of the new Ryazan City Committee for the Protection of Nature.

She established an environmental monitoring lab, procured a computer, and assembled a staff, which by 1993 had grown to fifty. Her committee, in collaboration with the Greens, developed environmental policy to propose to the Ryazan city authorities. The Greens focused on public education, while the committee addressed legislative and technical issues, such as the assessment and cleanup of contaminants from local petrochemical, leather-tanning, and electronic industries.

Artyomkina believes that citizens struggling for social change need to be empowered economically as well as politically. With that in mind she provided funding to a talented local scientist to help him develop pollution treatment technologies. He in turn pledged to assist her ecology committee in obtaining more monitoring equipment once his venture turned a profit. Artyomkina considers such entrepreneurship evidence that her country will recover in spite of recent economic setbacks. She believes that when Russians learn to live with their freedom the country will thrive.

However, Russia's economic disintegration in the 1990s provoked a backlash against the Greens, who were blamed for the closings of hundreds of enterprises for ecological reasons. Even though perestroika, the Soviet economic- and political-reform program, had lifted bans on publishing information about the state of the natural environment, "other prohibitions remained," Artyomkina says. "There were secret restrictions," nothing in writing, "but it was necessary to carefully decide how advantageous it was to speak, and to whom.

"A new class appeared—the mafia, corrupt bureaucrats who didn't want there to be open information about their acts," such as felling trees not authorized for timbering, or digging canals in protected areas.

> The mafia itself inflicted punishment for publishing such information. . . . When the Ryazan City Committee for the Protection of Nature, which I directed at the time, came to the defense of the Solotchin Forest and to the defense of land remaining free of development, it was objectionable to the mafia. So when our committee didn't give official permission [for these projects], the mafia itself in the person of the mayor of the city (my former comrade-in-arms in the struggle) . . . put forth lies and grave accusations against me, threatening fifteen or so years in prison in the best case, execution in the worst. They put me under investigation and this accusation became the pretense to dissolve our committee. The Green Movement organized in my defense, saved my honor and my life.

The committee, under Artyomkina's leadership, had become too successful in forcing bureaucrats to carry out its tough nature-protection decisions. "But the power of the mafia became stronger. Ecology affected too greatly the interests of the business world. . . . Many of our Greens who, at the beginning of perestroika, came to work in ecology professionally, could not hold their ground in their bureaucratic chairs. [They] became objectionable to the administration only for fulfilling their duties, and were squeezed out of their jobs and replaced by non–environmentally thinking powers."

When the Ryazan branch of the Russian Green Party protested the removal of the director of the local radio and television station, the former mayor of the city, in the presence of witnesses at a governor's reception, told the station's director that "if he didn't go willingly, then he would receive a bullet. Mr. Ryumen [the former mayor], now a rich man, switched from being mayor to banker, simultaneously the chair of a construction company and many other companies, the owner of which he became while still in government service."

⁂ Today Artyomkina leads the Ryazan branch of the Russian Socio-Ecological Union and is co-chairwoman of the Russian Green Party. "The base of my professional activity is publishing and printing . . . tied to the ecological movement and to the movement in opposition to the existing Yeltsin regime, which in our judgment is made up of mafia and bureaucrats." Publicly opposing Yeltsin, she says, has become dangerous. Nonetheless, she supported Mikhail Gorbachev in the 1996 presidential election.

She and her colleagues have founded a newspaper, *Green Beam*, and an independent printing house. They have also launched a project to create a public chemistry lab for NGO monitoring of environmental pollution. Artyomkina says that when the Ryazan City Committee for Nature Protection was dissolved, neither she nor her colleagues "left behind our ecological daring in that form in which it is possible today. . . . And perhaps our present activity in the social field is more useful and effective than activity within the government departments."

Artyomkina says she fears the ecological movement in Russia is in decline

and the green movement is suffering a deep crisis. To take part in ecology now is not only difficult but also dangerous. For environmental actions it is possible to pay very seriously, depending on whose interests you tread on and on what scale. . . . The terrible plunder of Russia's natural wealth goes on. People are more concerned about employment and increased salaries than [about] clean air and water. Now everything is possible—to talk, to steal, to consume, to slash, to destroy. . . . It is forbidden for one to be a [true] citizen of our country, to look after it, take care of it and augment its wealth. It is necessary [instead] to live by the laws of the ruling band of thieves, by their unwritten laws.

Artyomkina believes resolving the planet's current ecological crisis requires that there be no borders. "It is impossible to limit the flow of a common river or the movement of a general airstream." She says there needs to be a transformation in human consciousness, but, she observes, "human thinking is very inert. Five percent of the earth's population uses twenty-five percent of the planet's resources. To pull up the level of life of Africans or Asians to the level of life of Americans or Germans would mean to doom the planet to lifelessness and humanity to a still swifter movement toward our own death." She also believes that ending war and embracing the concept and practice of sustainable development must take place before global ecological problems can be resolved. She sees women as playing key roles in peacekeeping.

❧ Tatyana Artyomkina, twice married unsuccessfully, lives with her adult son. Her work schedule, plus her husbands' conceptions about a woman's family role, contributed to the failed marriages, she believes. The stresses of public work on personal life are exacerbated by the dangers involved in exposing corrupt practices in a corrupt system. Disregarding personal safety and ignoring threats from the KGB, Artyomkina spoke out in the early 1980s, going public with confidential information on health hazards caused by industrial pollution. She believed the public had the right to know. Fired from her government job because of her activism, she has become a Green leader, working to help her country recover from ecocide.

MARIA CHERKASOVA (1938–)
Unearths Health Facts

Another Russian environmental leader, Dr. Maria Cherkasova, maintains that—factoring in the cost of contaminated water unfit even for commercial use, much less for drinking, plus the value of cropland made unusable, plus working days lost to illness—ecocide had been depleting the country's strength and resources by 43 billion rubles per year (at the official rate about $26 billion U.S.). She observes that "[e]cological destruction has become a formidable barrier on the road to the economic and social rebirth of Russia."[3] Cherkasova believes Russia's ecological and economic problems are so closely interrelated that neither can be solved independently of the other—and she has the qualifications to make her case.

Maria Cherkasova—journalist, and biologist/zoologist/ecologist/ornithologist—since 1990 has been director of the Center for Independent Ecological Programs (CIEP). The group focuses on the problem of Russian ecological degradation as it relates to human health, especially that of children; she says the state pays too little attention to this issue. Yet environmental health problems are frightening. In one city, Archangel'sk, with a population of more than four hundred thousand, ten healthy women with healthy children could not be found to participate in a study.

Combining research with practical programs, CIEP brings together biologists, chemists, and physicians "working to protect the right of people to live in a healthy environment." They conduct research to identify the various toxic contaminants responsible for health problems. "We pay special attention to the influence of such super-ecotoxicants as dioxins."

In 1995 she created within CIEP the Ecology and Health Association, made up of parents with disabled children, health specialists, and representatives of advocacy groups, rehabilitation centers, and NGOs. This organization has become a vital tool in CIEP's work, Cherkasova says. It organizes seminars to educate persons about the most urgent environmental health problems, and it has just begun to publish educational literature on the subject for the public.

Married, with two children of her own, Cherkasova's special concern about the plight of Russian children led her in the early 1990s, in conjunction with some teachers and students at Moscow State University, to create (under the aegis of CIEP) the Suzdal Children's Center. Located in a 1500-square-kilometer park 200 kilometers from Moscow, the center provides a place where children with environmentally related health problems can be treated and rehabilitated in unpolluted surroundings.

"The main problem is ecological diseases connected with the anthropogenic environmental contamination, especially of [the] young generation. It's terrible, but when Russian children come to school there are almost no healthy ones among them." Cherkasova, founding director of the Suzdal Children's Center, convened an international workshop in Suzdal in June 1996 titled Children, Democratic Participation, and Community-Based Care. Her second international workshop there was Children and Sustainable Development. The center is supported by CIEP, the local Suzdal school administration, NGOs, and several groups in Moscow.

In 1992, this project earned for Cherkasova the esteemed Global 500 Award conferred by the United Nations Environment Programme. And in 1993 the Suzdal project received first prize at the Ford European Conservation Awards ceremony.

Cherkasova's most persistent and difficult obstacle is the lack of program funds for her center. Always under the gun to raise money, she notes that the concept of charitable giving is still in the embryonic stage, especially for ecological efforts, and giving hovers "near zero in Russia. We work due to the support of different Western foundations, mostly American."

The CIEP is a branch of the Russian Socio-Ecological Union, a countrywide coalition that Cherkasova helped organize in the mid-1980s and that she served as an officer. The largest NGO in the former Soviet Union, it includes over two hundred nongovernmental organizations in eighty-nine cities and eleven republics, each dedicated to defending the people's right to a wholesome environment.

Between 1988 and 1990, before she became director of CIEP, Cherkasova coordinated the Socio-Ecological Union's successful protest campaign against an ecologically destructive dam on the Katun' River, a tributary of the Ob' River, in central Asia. Born in Krasnoyarsk, Cherkasova had witnessed the demise of the beautiful Yenisei River near her home because of two huge dams; she vowed to prevent the same fate for the Katun'. "Due to the public efforts, this terrible dam wasn't built, though the danger is still real."

Cherkasova's commitment to a healthy environment dates back to the 1960s. As a student, she helped organize an amateur nature-protection club at Moscow State University that conducted studies and projects in the field. "We worked at the problems of nature protection practically, not in theory," she says. "[It was] important for developing independence, active civic position, [and] having a taste for scientific researches. In conditions of rough totalitarian regime of the Soviet Union putting down any forms of civic activity, it was especially valuable and it defined the whole life not only for me but many of my colleagues." The student nature-protection movement at Moscow State University that Cherkasova helped initiate continues its activities today. She remains in contact with the group and helps them as she can.

In 1990 Cherkasova came to the United States to set up an information exchange program with American environmental NGOs, learn about citizen protest strategies, and develop plans for working jointly with U.S. groups, including training in research and environmental analysis. During her visit she stressed that environmental issues cannot be separated country by country, but are common to the entire planet. "Your dirt is our dirt, and our dirt is your dirt. If America sends the Soviet Union polluting technology because you can make a quick profit from the sale, the wastes will return to harm you over time."[4]

Cherkasova, who as a biological scientist specializes in rare and endangered species, contends that the human population in Russia, at least in some regions, is dying out.[5] She calls it genocide by ecocide. Three of four pregnant women have health problems, and the mortality rate now exceeds the birth rate by a large margin. For one thing, the aftereffects of Chernobyl have been grossly underestimated. The Hiroshima bombing, for example, released only one-tenth the radioactivity of the Chernobyl accident.[6] Understandably, Cherkasova believes the development of nuclear energy was a tragic mistake and that it should be abandoned in favor of pursuing alternative ways of meeting energy needs.

Cherkasova reports that since 1949 the government has conducted 714 nuclear tests. And three nuclear disasters occurred in the southern Urals, one of which could be compared to Chernobyl. That explosion has been characterized as humankind's worst catastrophe, the effects of which in future decades will lead to many deaths. In Ukraine every square inch is contaminated with radiation. People's hair falls out, and blood runs from their noses. Even nuclear plants that never blow up pose dangers, polluting the water and causing sickness in nearby residents.[7] Accidents happen regularly in heavily populated regions—the areas usually selected for siting nuclear power plants. Under such circumstances, there is no safe place to live, Cherkasova says.

She points out that before 1987, even though she was never directly threatened by the authorities for her activities, the political oppression made it impossible for her even to consider creating her own organization or writing freely. But with the political changes since 1987, she says she has had "the opportunity to do what I consider vital and necessary and write about it. . . . Now the main direction of my work is the problem of human beings as rare and endangered species and defense of their right for life; it concerns especially children."

Cherkasova writes a regular ecological column for a popular science journal in Russia. In the early 1990s she invited readers to respond to an environmental survey she published. Letters poured in, more mail than she had ever before received in her twenty years of writing for the journal. The letters contained tragic stories of persons suffering from environmental pollution and the effects of radiation from nuclear accidents and tests. Typically they were kept in the dark about what had happened. Cherkasova quotes from some of these letters in an article she wrote for the March 1992 issue of the *UNESCO Courier*. Here is one example.

On 29 September 1957, almost thirty years before Chernobyl, an explosion took place at Chelyabinsk-65, a secret city in the southern Urals where nuclear waste was stored in a chemical complex. The blast shot radioactive debris more than a thousand feet into the air, polluting nine thousand square miles.[8] Almost twenty million curies of radiation were released. Zoïa Islamova, a woman in that region, wrote to Cherkasova,

> Nobody told us what had blown up, and we were not allowed to discuss it. The neighboring villages were evacuated. People who refused to leave were forcibly expelled and their houses were burned down, even though setting fire to the buildings was the last thing that should have been done. We learned exactly what had happened only in 1989, when it began to be talked about in the newspapers and television. The rumor was that they were going to evacuate our town too, but that would have cost the state so much that they decided to leave us where we were, like laboratory guinea pigs.

Zoïa fell seriously ill not long after the accident. Her husband, who was also affected, is now an invalid. "There is a lot of sickness in our town," she went on,

> bronchial asthma, allergies, skin conditions, problems with joints and the digestive tract. We were never measured for levels of radioactive exposure, and doctors refuse to make the connection with the 1957 explosion. But we all think it is the cause of our problems. We no longer trust anyone or anything. Life is terrifying.

In her article, Cherkasova reveals that "[m]any letters came from the Volga area. What was once Russia's great river is now disfigured by gigantic dams and transformed into a sewer for military and industrial wastes. . . . Every letter described the same ecological problems and told the grim story of misery and despair. Only the name of the town or village changes."[9]

In answer to her own question, What can we women do in the face of such a threat? Cherkasova declares,

> First of all we must do all we can to sound the alarm and wake the dormant survival instinct of the human race, first and foremost among the representatives of the stronger sex that hold power, and yet seem singularly powerless. . . . We must demand that the country move immediately to a path of peaceful development, and reject colossal military expenditure, atomic bombs and other even more sophisticated weapons, the arms trade, the priority given to heavy industry ever since Bolshevik days, and huge, destructive projects. . . . And why not a Women's Parliament? to defend the rights of women and children and to propose an alternative, crucial for humanity's future, to the male military and technocratic model that has brought the world to the edge of the abyss.[10]

Internationally acclaimed as a journalist and a scientist, Cherkasova has carried out her crusade through her writings, lectures, and leadership as head of an NGO within the Russian Socio-Ecological Union. During the past two decades she has written over a hundred articles and books—some about rare and endangered species, others exposing suppressed facts about environmental degradation and its effects on human health, especially the health of children, in the former USSR. For example, in the mid-1990s Cherkasova presented a paper entitled "Children and Ecologically Related Diseases in Russia" (later published) at an international conference, Children and Environment, held at Norway's Center for Child Research. She has written a series of popular scientific books, *They Must Survive*, about rare and endangered animals. In October 1996 she participated as a featured speaker at a Childwatch International Key Institutions Workshop in the United States.

MARIA GUMIŃSKA (1928–1998)
Kraków's Pollution Buster

The countries of the old Warsaw Pact belong neither with the rich nations of the North nor with the poor nations of the South. Nevertheless, their peoples have valuable knowledge of the hazards of forced industrialization. In central and eastern Europe, the trauma of Nazi tyranny and the horrors of Auschwitz presaged nearly fifty years of another kind of repression. The Soviet military-industrial complex controlling these countries following World War II devastated their natural environments. In Poland, one of those hardest hit by industrial pollution, air quality became so bad in some areas that people took refuge in salt mines.[11] The death rate in some parts of the country is still higher than the birth rate.

The medieval city of Kraków, site of a remarkable historical complex, was once the nation's cultural center and was its capital from 1058 to 1609. During the past half century, however, Kraków has suffered severe environmental degradation from a variety of heavy industry complexes. Air pollution from intensive coal burning darkened the sky and assaulted inhabitants' lungs. In Skawina, only twenty kilometers from the heart of the city, Communist industrialists had erected an aluminum smelter emitting toxic fluoride. A suburban steel mill added thousands of tons of carbon monoxide and heavy metals and other carcinogenic pollutants to the air. Within the residential area of the city proper, a chemical/pharmaceutical factory emitted toxic chlorinated hydrocarbons. As a result, Kraków, Poland's third-largest city, became one of the most polluted in Europe. Placentas examined after delivery of "healthy" babies in Kraków showed high concentrations of fluorine and such heavy metals as lead, mercury, and cadmium.[12]

Dr. Maria Gumińska, until her death in 1998, lived in that residential neighborhood in Kraków. Physician, biochemist, professor, researcher, NGO leader, prolific writer, and activist, Gumińska said, "A political decision was made to change the social structure of local society by eroding the intellectual class—which was largely in opposition to the ideology being introduced by communist authorities—through an influx of a 'revolutionary' workers' class. After some years, the first signs of ecological disaster were visible, but the information on environmental issues was kept top secret by the government."[13]

For Gumińska, who had grown up in an unpolluted and beautiful environment in the Tatra Mountains, near Poland's southern border, the air in Kraków was a violation of her senses. She recalled in the mid-1970s being "so sensitive that I could feel in the central part of Kraków the characteristic, irritant smell of hydrogen fluoride in the ambient air. I remember one evening, when I was walking with my husband in the city, I understood that it was the beginning of extermination of the whole area." Gumińska's husband, a physician and pediatrics professor, developed cancer a few years later. She believed he was a victim of exposure to chlorinated carcinogens from the chemical factory near their apartment. "For many years we had a desert environment in our neighborhood. At night in our apartment we were like in a gas chamber."

In 1969 Gumińska had been assigned to investigate fluorine toxicology with a group of specialists associated with the Kraków branch of the Polish Academy of Sciences. She conducted health studies on workers in the electrolysis division of the aluminum smelter who were exposed to fluorides. "In 1970, I was shocked because the workers, after ten to eighteen years of work in the smelter, had become so weak they could not straighten their backs to a normal vertical position. They suffered from spine pain and all of them had osteofluorisis.[14] I discovered that the weakness was caused not only by degenerative changes in their skeletons, but also by a decrease in chemical energy formation and content in their organisms."[15]

Gumińska and her fellow investigators tried to publish their data in 1975, but the authorities forbade the material's release. By 1978, her anxiety about pollution had become intense. "It was not only a 'Silent Spring' but a 'Dead Spring.' In the vicinity of my home many trees died, they had no leaves, the result of toxic influence. . . . With my 'chemical nose' I could feel some chlorinated hydrocarbons in the air. Our situation, as inhabitants of the area, was really hopeless. I had nothing to lose and I decided to fight against it."

During the rise of the Solidarity movement in the autumn of 1980, Gumińska and like-minded citizens and ecologists—many of them women concerned that the city had been reduced to an ecological disaster—founded the Polish Ecological Club, or PKE, the first independent ecological NGO in Poland. Made up of journalists, scientists, physicians, and engineers, the PKE actively confronted Communist policy, which disregarded human health, nature, and historical sites.[16]

About that same time, in December 1980, Gumińska had her article about fluoride poisoning from the aluminum plant published in the local newspaper. Ironically, during the period when the fluoride poisoning of Kraków was coming to light, a government commission in Warsaw had sent an emissary to the city to discuss community water fluoridation to prevent tooth decay. In the face of the massive toxic emission of fluorides from the aluminum works, the consequent contamination of the environment, and damage to human health, however, the emissary quietly retreated to the capital without broaching the subject.

Shortly thereafter the city authorities requested a report from Kraków's ecologists on the health hazards of fluoride. Within two months after receiving their report, Kraków's city president and the central Polish authorities shut down the electrolysis department of the aluminum smelter, the source of the fluoride contamination.

"This was a great achievement," Gumińska wrote, "because for the first time human health and social problems had been taken into consideration. The factory continued aluminum production by recycling aluminum waste."

Unfortunately, fluoride had reached some local drinking-water sources. Solid waste from the aluminum works containing soluble fluorides had been used in surfacing roads in the region. Rainfall runoff carried the fluorides into the groundwater and into wells, and also contaminated fifteen thousand hectares of soil. Tests revealed the fluoride in the water exceeded the maximum permissible concentration by one hundred times. "This explained the dramatic changes in the health of people exposed to fluorides not only from polluted air, but also from the drinking water. So it became necessary to construct new water pipes which supplied the area with clean water transported from a long distance, and to recultivate contaminated soil with calcium carbonate or dolomites."

The unilateral decision of local authorities to protect their citizens by closing down part of the aluminum plant led to the city president's being placed under Communist surveillance in the spring of 1981.[17] To stop such uncontrolled actions in the future, the Polish authorities declared martial law on 13 December 1981 and subsequently arrested many people.[18]

The day before, Gumińska had been returning by car to Kraków from a district hospital some 130 kilometers away, where she had lectured to doctors on the treatment of fluoride poisoning. Prompted by intuition, she instructed the driver of the courtesy car provided by the hospital to take a detour off the main road, which was crowded with military vehicles. As the driver turned right onto the side road, they heard a disturbing noise on Gumińska's side of the car. Stopping immediately, the horrified young driver discovered that the metal parts holding the right front wheel

COURTESY MARIA GUMIŃSKA

MARIA GUMIŃSKA AT HER DESK IN KRAKÓW.

in place had been severed save for a thin thread. Polish vehicles are required to carry spare wheels and the tools for changing them, so he was able—though shakily—to make the necessary repair.

The declaration of martial law suspended the permits of citizens' groups, making it unlawful for them to meet or engage in actions. Nevertheless, since the founders of the fledgling Polish Ecological Club had not yet officially registered their group, its existence remained unknown to the authorities, and they were able to continue their activities underground. Martial law was ended in 1983.

Gumińska's leadership role in the PKE kept her in the forefront of Poland's environmental movement. She served on the group's board of directors for three years. She worked in coordination with the national PKE, headquartered in Kraków, and led its health commission, which addresses ecological as well as human health issues. "We succeeded," she wrote, "in our fight against a local chemical factory, which had to change its harmful technology as a result of social pressure and stop emitting mutagenic chlorinated hydrocarbons which are harmful for plants, animals and human health. The change in technology actually helped the economic situation of the factory and liberated Kraków from toxic chemicals."[19] Furthermore, after years of negotiations with the nearby steel mill, Gumińska and her PKE colleagues achieved by the early 1990s a radical reduction in emission pollution, dropping it to 30 percent of what it had been in 1980. For social and economic reasons, however, the heavy concentration of obsolescent industrial facilities in Kraków cannot be eliminated quickly. Nevertheless, improvements are already apparent, especially in the restored historical areas. Finally, the PKE is striving to educate authorities and citizens about the advantages of alternative energy sources and public transit. The NGO is also pushing to outlaw leaded gas. Overall, the city has made noticeable strides in becoming less gray and smoggy.

Under the auspices of the PKE, Gumińska organized a one-day seminar in Kraków in March 1995, The Role of Women in Environmental Protection, during which a group of women environmental activists from Japan shared experiences and expertise with their Polish counterparts. The PKE published the proceedings of the seminar. Also in 1995, Gumińska represented the PKE at an international meeting of NGOs in

Manila, Philippines. Participants hope to organize a global network of groups working together on environmental and social problems. She believed such international citizen action was the route to solving Earth's present ecological predicament.

Women make up more than half of the Polish Ecological Club; in the most polluted regions of the country, they often lead the ecological organizations. The PKE now has 4,000 members, 15 branches, and 120 circles throughout Poland. It organizes courses, lectures, meetings, ecological issue campaigns, discussions, and seminars. The club convened a conference on sustainable development (one of its first seminars) in 1985 and published the proceedings. Members of the PKE undertake numerous action projects—such as those involving the reduction and management of waste—to help restore the environment. Another goal is to introduce ecological education at all levels in schools and make it obligatory. The organization publishes books and literature to develop environmental awareness, which is low among ordinary Poles. The PKE puts out a regular newsletter, *To Be or Not to Be*, as well as materials for teachers and parents—leaflets, posters, booklets, handbooks, and the proceedings of its numerous conferences. The club recently published in Polish a series of eight books entitled *To Save the World*; now translated, it is available in seven languages.

Maria Gumińska believed that Poland has ideal conditions for developing organic farming, and the Polish Ecological Club and other NGOs lobby for it. However, the government continues to promote energy- and chemical-intensive industrial farming, which leads to soil degeneration. The PKE has trained six hundred farmers (mostly women) in organic agriculture. Some regional branches of the club enter into cooperative arrangements with organic farmers to distribute healthy vegetables through a network of shops. One such PKE branch in the Silesian region has twenty-four vegetable shops in twelve towns. Of the eighteen local groups of the organization in Silesia, twelve are led by women. The PKE works internationally as a member organization of the World Conservation Union, Friends of the Earth International, Environment Liaison Centre International, and the Central and Eastern Europe Bankwatch Network (composed of NGOs and banks).

Gumińska worried that with liberalization eastern Europeans eager for the lifestyle standards of the West and influenced by its print and broadcast advertising—along with a flood of enticing, overpackaged consumer items—would repeat the ecological mistakes of Western countries rather than learning from them and choosing a wiser course.

Kraków's historic city center has been inundated with commercialization and advertising alien to its medieval character. Its historical identity is threatened by international corporations like McDonald's, which is seeking a site in the protected historic district of the Main Square, whose design dates back to the eleventh century.

Gumińska, widowed in 1985, continued until her death to teach biochemistry at the medical college of Jagiellonian University, one of the oldest universities in Europe. The Chemistry Faculty recently established environmental science as a new specialization. Gumińska lectured on the medical aspects of the subject to students enrolled for that major.

Through her many publications (close to a hundred), Gumińska popularized ecological issues among social organizations and in the scientific community. What kept her going in the face of the enormity of the environmental degradation that eastern Europe suffered in this century? She said that everyone has a moral obligation to protect the resources of the earth. "The Earth is not a gift from our ancestors, but a loan from our grandchildren. Until 1980 we were not able to do anything. Later we were happy with our small achievements and some improvements; but now we again see new problems connected with consumptive lifestyles and ethics." She said women employ a different approach than men in solving ethical, ecological, and economic problems. "They should have a stronger influence on political decisions, and they should play a dominating role, not in imitation of men, but in mitigating their behaviors."

CHAPTER EIGHT

*Perseverance is more prevailing than violence; and many
things which cannot be overcome when they are together,
yield themselves up when taken little by little.*

PLUTARCH
The Parallel Lives

Perseverance and Patience Pay Off

CAROL BROWNER (1956–)
Hangs Tough

In a meeting room at the Orlando airport early in 1992, Carol Browner faced half a dozen lawyers seeking fast approval on a permit to expand the airport. Browner refused to approve the proposal as drafted because it did not comply with state regulations. A lawyer herself, she cited the requirements that the expansion request did not meet. The lawyers argued on for several hours, even becoming verbally abusive when she refused to yield.

As secretary of Florida's Department of Environmental Regulation (DER) for two years in the early 1990s, Browner had streamlined the state's regulations to shorten the review time for permit applications. Nevertheless, she insisted that the regulations be followed.

As a native of Florida who had grown up near the Everglades, Carol Browner well understood the importance of tourism to the state's economy. So as head of the DER she faced the dilemma of accommodating tourism without destroying Florida's fragile ecosystems, especially its wetlands. With Disney World about to undertake a major expansion, she appreciated the need for an enlarged airport. By persevering all afternoon, however, Browner prevailed, and the airport lawyers agreed to comply with everything she demanded.

Browner also negotiated an agreement with Walt Disney World Company whereby, in return for state and federal permits to fill in 400 acres of wetland for expanding its operation, the company agreed to commit $40 million to purchase an 8,500-acre ranch, restore it as a wetlands area, and donate it to the Nature Conservancy as a wildlife refuge.

Browner, tall, slender, and only thirty-five years old at the time, did not imagine she would soon be recruited to become the first woman administrator of the U.S. Environmental Protection Agency (EPA).

Washington looked askance at Browner as she moved through the confirmation process. And once she had taken over as head of the EPA, questions like Is Carol Browner in over her head? floated around the business, environmental, and government communities.

After settling into the seedy EPA headquarters in a rough section of Washington, she quickly demonstrated her mettle. Smart, pragmatic, and a rapid learner, Browner had acquired a mix of training and experience qualifying her for the demands of running the large regulatory agency. She had worked eight years in Washington after law school, first as an environmental lawyer for a nonprofit consumer group, then as a staffer for then Senators Lawton Chiles and Al Gore—both committed environmentalists. No one else who had ever headed the EPA brought to the job both congressional and regulatory-enforcement experience.

When Chiles became governor of Florida in 1991, he brought Browner into his cabinet as secretary of Florida's DER—one of the nation's largest environmental regulatory agencies. During her brief tenure, she earned praise for her patient and effective handling of difficult and complex issues involving restoration of the Everglades, hazardous-waste disposal, and wetlands protection. She also succeeded in getting a gas-tax increase through the legislature to pay for cleaning up the environment.

Following the 1992 national election, Browner—in addition to coping with the aftermath of Hurricane Andrew in Florida—had taken on the assignment of helping vice president–elect Gore with the new administration's transition process. The subsequent interview with president-elect Clinton for the EPA post came as a surprise. She had been happy in her Florida job and was sorry to leave it.

The legislation creating the Environmental Protection Agency was approved by Congress in December 1970. The new agency's mandate was to solve the nation's urgent environmental problems and protect public health. It enforces such environmental legislation as the Clean Air and Clean Water Acts, the Resource Conservation and Recovery Act, and the Superfund law, as well as establishing rules for pesticide use and food safety. Browner observes: "Some of the problems from 1970 have proven to be difficult to solve. Also, by learning more about the environment, we have discovered new problems that must be addressed in new ways. . . . We at EPA enter our second quarter-century committed to providing a cleaner environment, at cheaper cost to the American economy, using smarter strategies of environmental protection."[1]

Browner inherited an agency criticized for poor management, rules violations, and inefficiency. After taking over she implemented internal reforms, as well as issuing a package of twenty-five rule revisions to reduce paperwork and simplify environmental rules.

Preventing pollution before it happens rather than waiting to clean it up at the end of the pipe heads Browner's priority list. At the EPA she began a new industry-by-industry approach to regulation. She has also initiated a more comprehensive procedure for dealing with environmental problems, developing integrated solutions for whole ecosystems and communities. "We make very bad decisions when we fail to look at how something functions within a system as a whole. . . . Moving from a permit-by-permit piecemeal approach to a holistic watershed approach is a difficult transition."[2]

Browner has expanded the Toxic Release Inventory, which, under the community right-to-know law, requires industry to make information on toxic emissions public. Under her leadership, the agency has adopted the first standards ever for incineration. "We've cut toxic air emissions by 90 percent; we've cleaned up more toxic waste sites in three years than in the first twelve years of the Superfund Toxic Waste Cleanup Program" (*AS*).

Browner is particularly concerned about wetlands protection, since she considers them our most important resource. "Wetlands give us our diversity of species: they are nature's kidney, a way for nature to clean up the pollution that we discharge, to keep wildlife happy and to protect us and our homes from erosion and flooding" (*AS*). In May 1996, she announced that the EPA had won in court a $22 million penalty for a wetlands violation case, the second-largest enforcement action in U.S. history under the Clean Water Act—exceeded only by the *Exxon Valdez* oil spill case. A pipeline company had bulldozed and contaminated a wetlands. The EPA enforcement action required the company not only to pay a civil and criminal penalty of $22 million but to restore the wetlands system.

Browner does not always win. For example, she lost a suit brought against her by Ethyl Corporation. She had denied them a permit to use a new fuel additive because no studies had been done on its health effects. The judge ruled against the EPA, holding that the law says health studies can be done once the product is on the market. Browner believes that common sense demands doing health effects studies at the front end. Twenty years ago, she notes in an interview with the *Christian Science Monitor*, when lead was banned from gasoline, there was not yet numeric certainty about costs and benefits. All we knew was that children were at risk. Instead of waiting ten years for data to be compiled, she says, we were proactive, and ten years later there was confirmation that the benefits justified the costs.[3]

Whether or not the EPA administrator should have veto authority over permit decisions involving the U.S. Army Corps of Engineers has been an ongoing controversy. The issue surfaced again in the early 1990s, when Browner's predecessor had exercised the agency's veto authority to stop the Corps of Engineers from constructing the Two Forks Dam on the Platte River. Since that action, the corps had been in court challenging EPA's veto authority. Browner and her colleagues at the agency, believing an important principle was at stake, followed the court's proceedings for over three years, arguing at hearings for the EPA to have such authority. In June 1996, the judge ruled in favor of the agency's right to say no to the Corps of Engineers where water quality, quantity, and biodiversity were put at risk.

The issue of the EPA's veto authority arose again during the budget process in the 104th Congress. Pressured by lobbyists from powerful oil, chemical, and pharmaceutical companies, Congress cut the agency's budget by 30 percent across the board, along with a 40 percent cut in its enforcement program. Congress also attached seventeen riders to the appropriations bill, each one allowing a special deal for a special interest.[4] One of the riders dealt with the EPA administrator's veto authority.

Browner, with the help of the environmental community, lobbied in defense of her agency's budget and against the riders through press interviews and in meetings with individual members of Congress. The dispute lasted for months. The agency had to shut down twice when the president, with Browner's support, vetoed the proposed budget. Eventually, six months into the new fiscal year, the budget bill passed with most of the money restored and all of the riders removed.

"When we began to fight against the riders, people didn't think I could win; but I can hang tough; I can stand firm, and we did win" (*AS*).

The senator who had attached the rider about the EPA's veto authority told Browner later that he intended to attach the same rider to the next budget proposal. He argued that there was duplication of authority between the EPA and the Corps of Engineers. Browner later recalled the rest of their exchange:

> And I said, "Well, I don't agree with you, but if you do believe there's a duplication, could you please tell me why you don't transfer all of it [authority] to EPA and away from the Corps?" I got only silence. I believe he thought I had crossed the line of being inappropriate. After all, he's allowed to question me, but I'm not allowed to question him. I said to him then, "I don't understand how you could say to the EPA and to the American people, the EPA is responsible for your water quality, for the health of your lakes, your rivers, your streams, and we're telling them to do that job; but we're telling them to do it with their hands tied behind their backs. We're not going to let them protect the wetlands that are essential to the health of your river and your lake and your stream." (*AS*)

Although the word "environment" did not appear in most published versions of the Republican House majority's *Contract with America*, Browner says that reading the fine print revealed that most of its ten proposals dealt with weakening environmental protection.

Members of the House of Representatives of the 104th Congress decided to rewrite the Clean Water Act, a law fundamental to maintaining a healthy environment. They called in people from the same special interests who had contributed to their political campaigns to draft the bill.

> I've worked in Washington on and off for fifteen years, and this was truly a remarkable event. They literally asked the special interests to come into a room—we were not allowed into the room, the people who would have to enforce the law were excluded from the discussion. They parceled out to each of the special interests the responsibility for rewriting a particular section of the Clean Water Act. Then they introduced that special interest piece of legislation to the House and passed it. It is a bill that would systematically weaken each and every tool we have used to clean up our water, to protect our rivers, our lakes, our wetlands, our beaches. Make no mistake about it, it would roll back twenty-five years of environmental progress. (*AS*)

The media dubbed the proposed bill the Polluters and Developers' Relief Act, or the Dirty Water Bill. When Browner told the president about the legislation, he said, even before the Senate had considered the bill, that he would use his veto. It never came up for discussion in the Senate.

Browner also had to battle the 104th Congress over the fifteen-year-old Superfund law. She maintains that serious problems with the law as originally written hamper its enforcement. While considering the changes Browner had proposed, Congress decided that the principle fundamental to the Superfund—that the polluter ought to pay—should be eliminated from the law, shifting cleanup costs to the public. Browner would not agree to that, so attempts to improve the Superfund law stalled.

Browner works to dispel the belief that cleaning up the environment hurts the economy. She has launched what she calls Common-Sense Initiatives, which save billions of dollars while maintaining standards. The EPA under Browner is compiling a series of reports demonstrating that what is good for human health and for wildlife is also good for business, that a healthy environment and a healthy economy go hand in hand. The agency has examined four or five economic sectors in preparing these reports. One, "Liquid Assets," released on Memorial Day weekend 1996, details the economic value of clean rivers, lakes, streams, and beaches.

"We found that the recreation and tourism industry, which is a $380 billion a year industry—the second-largest employer in the United States—is almost absolutely dependent on clean water: 1.8 billion trips will be taken to rivers, lakes, and beaches in the United States this year. It is our favorite place to recreate and that industry—and I know it's true in my home state of Florida—is absolutely dependent on clean water" (*AS*). The EPA prepared a similar report on clean air, substantiating that benefits surpass costs.

Browner and the EPA, working with federal and Florida agencies and with the Everglades Coalition of environmental organizations, proposed a comprehensive federal action plan for restoring the Everglades. "For the first time ever we are bringing together all the pieces of the federal government, each of us doing our part to restore the heart of the Everglades—to see it once again pulse with the water that is essential to its health, its future" (*AS*).

Staff members of the U.S. Fish and Wildlife Service (USFWS) in the Department of the Interior complain that the EPA does not pay enough attention to wildlife, focusing instead on hazards to people. They claim that many pesticides registered in the United States, such as phorates, aldicarb, parathion, carbofuron, and diazinon—even when used as directed—are killing sixty-seven million of the country's birds every year and should be banned because there are safe alternatives. Letters and memos from the Fish and Wildlife Service to the EPA requesting action have proven futile; the USFWS also points out that the agency, in response to requests from agribusiness, issues numerous exemptions to restrictions and extends permits almost without question. They contend that pesticide regulators at the EPA favor pesticide manufacturers, often going to work for them after leaving the agency. Under the Federal Insecticide, Fungicide, and Rodenticide Act, which the EPA administers, killing birds with pesticides, even accidentally, is illegal. But the USFWS staff maintain that EPA is not enforcing that law.[5]

Browner agrees that the law governing chemicals, pesticides, fertilizers, herbicides, and fungicides (passed more than forty years ago) needs revising. She submitted a revision proposal to the 104th Congress, but it did not pass. The obsolete law focused mainly on carcinogens that

EPA's position as an adjunct to government, making it an integral part instead. Others have hailed Browner as a new breed of ecologist.

JANET GIBSON (1953–)
Creates Central America's First Coral Reef Reserve

Belize (known as British Honduras until 1981), with a population of only two hundred thousand, boasts one of the colorful jewels of the sea: a coral reef that is the second largest in the world and the longest in the Western Hemisphere. Squeezed between Mexico's Yucatán Peninsula and Guatemala along the eastern coast of the Central American isthmus, tropical Belize has an extensive marine ecosystem, including the 137-mile-long reef, which marine biologists consider one of the seven underwater wonders of the world.

In the mid-1980s, Janet Gibson, a third-generation Belizean, became concerned about the threat to the coral reef from increasing development pressures: overfishing and the use of dynamite and cyanide by fishermen; dumping industrial wastes and sewage into the Caribbean; using beach sand and coral for construction; unmanaged tourism, bringing with it careless boat anchoring and divers trampling the coral; and dredging and filling operations. She had reason to worry. Coral reefs are fragile. Even climate change damages them; it takes only a slight increase in sea temperature to ruin 90 percent of a coral reef.[7]

Rivaled only by tropical rain forests in the diversity of living things for which they provide habitat, coral reefs have enormous economic value to countries around the world. And yet they are being destroyed by human activities at a rate of over 230 square yards per second.[8] According to a 1987 survey conducted by the World Conservation Union and the United Nations Environment Programme, 93 of the 109 countries having coral reefs of significance are damaging these rich biological resources.[9] Scientists report that one-tenth of the world's coral ecosystems have already been lost, and an additional two-thirds may face ruin.[10] Coral reefs date back five hundred million years, making them among the planet's most ancient ecosystems. They depend on sunlight and flourish best in clear, shallow coastal areas. Although they look like they are made of rock, the reefs actually consist of living organisms called polyps. Regeneration of damaged coral reefs takes decades. A number of countries have made it illegal to collect coral—either dead or alive.

Even though coral reefs are hidden under the ocean's surface, out of immediate sight, they benefit humans in countless ways. These complex ecosystems, occupying only a fraction of 1 percent of the earth's surface, provide homes and breeding areas for 25 percent of all marine species.[11] They shelter shorelines from storms and erosion, create sandy beaches, and produce chemicals having medical applications. Fish populations depend on ecosystems supported by coral reefs, making possible the production of high-quality protein from nothing but sea water—essentially zero-cost, self-sustaining fish farms.

Janet Gibson, biologist and zoologist, grew up near Belize City close by the sea, visiting the coral islands (cayes) with her family, including five siblings, during summer holidays. She recalls, "We had no television here then, so we spent a lot of time outdoors, on the beach and knocking around in boats. I always had an affinity for the sea."

Gibson became aware of the exploitation of coral reefs by the fishing and tourism industries while working for the Belize Fisheries Department in the mid-1980s, a job she had obtained after earning her bachelor's degree in botany and zoology at King's College of the University of London. Gibson knew that a healthy barrier reef was central to Belize's economy.

Concurrent with her Fisheries Department job, she worked part-time for the Wildlife Conservation Society (WCS) and as a volunteer with the then fledgling Belize Audubon Society. Visiting the village of San Pedro on Ambergris Caye, the northernmost coral island in the reef chain, where the WCS had a project under way on groupers, Gibson remembers that

> [i]t was clear that the community was interested in some sort of protective measure for the area. San Pedro had changed from being a sleepy fishing village to quite a bustling tourism center. The significant increase in diving, boating, and snorkeling on the reef was a cause for concern as the reef was being impacted. I knew that the fisheries legislation provided for marine reserves, but none had ever been declared. So, with encouragement

from Belize Audubon and WCS I decided to do a management plan to set aside five square miles of reef area as the Hol Chan Marine Reserve. [*Hol Chan* means "Little Channel" in Mayan.]

Selling the concept of long-range strategic planning—a new idea for Belizeans—meant patiently educating government agencies, commercial interests, and the public to its advantages. At first the fishermen resisted the idea, and commercial interests worried about negative economic consequences. But Gibson persevered, refusing to accept defeat. She says of the experience:

> The most difficult time was getting the fishermen to endorse the concept of a protected area where fishing would not be permitted. Fortunately, most of the fishermen were members of a cooperative in San Pedro, so at least it was easier logistically to talk to them. Gradually, most of the fishermen became supportive of the idea. Their reluctance was only natural as we were seen to be "taking away" part of their fishing grounds, although only a handful of fishermen were actually affected. I took part in some rather fiery meetings from time to time, and I particularly remember one occasion when an uncivil fisherman's remarks almost reduced me to tears. But I persevered, trying to focus on the positive impacts that would result in the long term. In the end, consensus was reached and the cooperative formally agreed to the Reserve.

In addition to persuading the fishermen, Gibson campaigned to win support from citizens and from persons involved in commercial tourism. At the same time she prepared the long-term strategic planning document for the Hol Chan Reserve, and helped secure financial support for its maintenance and administration. In 1987, after two years of intensive effort, the reserve became a reality. With contributions from environmental NGOs, the Belize Fisheries Department purchased a patrol boat and established an educational center.

Since the Hol Chan Reserve was established, the number of commercially valuable fish nearby has increased, and the appeal of an almost pristine coral reef environment has produced a thriving tourist business, now managed to protect it. As of 1996, a proposal to enlarge the five-square-mile Hol Chan Reserve was under consideration. But even more important, Gibson says, other communities in Belize are planning and creating similar reserves. For example, she helped establish Belize's Glover Reef Atoll as a protected marine reserve. The Goldman Environmental Foundation selected her in 1990 to receive one of its coveted prizes as an environmental hero.

Experts believe that 10 percent of the known species in coral reefs will become extinct in the next decade; moreover, as many as 90 percent of the species in reefs still remain undiscovered.[12] Global concern for the loss of coral reefs, growing out of the 1992 Earth Summit, led to the establishment of the International Coral Reef Initiative in December 1994 to help protect coral ecosystems and to provide better decision making through a science-based information network. The group launching this initiative, which includes the United Nations and the World Bank, proclaimed 1997 the International Year of the Reef. Over a hundred governments and NGOs joined this effort to rouse public awareness.

Gibson not only introduced the concept of long-term strategic planning to Belize but she demonstrated how to put it into practice. Widowed in 1990, she continues to lead in protecting Belize's marine ecosystems. Gibson is now at work on her country's Coastal Zone Management Project to expand and strengthen the program to ensure sustainable use of marine resources and to protect their unique diversity. In 1995 she obtained her master's degree in tropical coastal management.

POLLY DYER (1920–)

Relentless Advocate for Washington State

One afternoon in 1956 Polly Dyer—a Mountaineers Club and Sierra Club activist, and president of the Federation of Western Outdoor Clubs—huddled in her Auburn, Washington, living room with two leaders of the Wilderness Society to brainstorm. What could they do to forestall plans by private interests to build a highway along the pristine coastal strip of the Olympic National Park—the longest roadless coastline in the continental United States? Even the Park's superintendent favored the highway.

What about persuading an associate justice of the Supreme Court, William O. Douglas—a Washington native—to lead a well-publicized three-day hike along the Olympic Park ocean coast? Perfect, they agreed; but would he do it? Earlier he had led a similar expedition along the Chesapeake and Ohio Canal to demonstrate that an expressway should not be allowed to replace the hiking path there.

Now recognized internationally as a World Heritage Site and a Biosphere Reserve, the Olympic National Park in northwestern Washington encompasses a large unmanipulated ecosystem, including a temperate rain forest, alpine forests, glacier-capped mountains, and the unique coastal strip. But in the 1950s no Wilderness Act existed, and the park had no such protected designation. Concern for environmental protection had not yet become widespread. Rachel Carson's *Silent Spring*, the Environmental Protection Agency, and modern environmental laws had not yet appeared. Nevertheless, a culture of conservation already existed among the individuals who had founded such early conservation organizations as the Sierra Club, the Wilderness Society, and the Audubon Societies. It existed, too, within regional groups like the Olympic Park Associates, begun in 1948 to protect the integrity and wildness of the park.

Enlisting the help of Justice Douglas in protesting the coastal highway plan appealed to Polly Dyer and her colleagues. He accepted their invitation, and Dyer agreed to organize and coordinate the event. The hike, sponsored by the Wilderness Society and the Federation of Western Outdoor Clubs (Dyer was its first woman president), took place in August 1958 and included seventy-two conservation leaders. "Subsequently," Dyer recalls,

> in 1964, Justice Douglas contacted the Olympic Park Associates proposing a reunion hike for the southern half of the Olympic National Park coastal strip we had not hiked in 1958, organized again by [me]. We included anyone who wanted to participate. This time there were 158. Ordinarily, none of us want to have large groups in wild, natural areas; but those two events, of course, had publicity as their purpose. . . . I will be forever convinced that the 1958 hike laid the road proposal to rest, although it kept resurfacing every once in a while; and the 1964 trek reinforced keeping it roadless.

Combatting further ecological impoverishment of Washington State, and repairing damage done by the extractive industries over many decades, have driven Dyer's volunteer activism for over forty years. She has been among the founders of six different environmental groups formed to focus on specific causes, and she has held continuing leadership roles in almost all of them. She now is active in at least ten organizations as an officer or board member. In the mid-1960s when the Dyers purchased a home in the Seattle area, she insisted on having a living room large enough to accommodate big environmental gatherings.

Dyer, who characterizes herself as a childhood introvert, comes from a family having little interest in the outdoors. What triggered her concern for matters environmental, especially her passion for protecting wilderness? She credits two experiences. First, as a young woman in 1940 she hiked to the top of three-thousand-foot Deer Mountain near Ketchikan, Alaska, after her family moved there. She was overwhelmed, "experiencing, for the first time, the extent of wild, open mountainous and forested areas." The second defining moment was a chance encounter. "Several years later, meeting John Dyer on Deer Mountain, I asked him what the Rock Climbing Sierra Club pin on his hat was all about. He had been an active leader in the San Francisco Bay Chapter before coming to Alaska; after we married several months later, the first thing I had to do was join the Sierra Club. His *Sierra Club Bulletins* were full of conservation, the beginning of my education."

After settling in Washington State, the Dyers, together with a friend, formed the first Sierra Club chapter outside California. Polly Dyer soon became the first Sierra Club national board member not from California, serving a six-year term. While living briefly in Boston, she and a friend started a New England chapter.

As a founder in 1957 of the North Cascades Conservation Council (conceived in her living room), Dyer pushed successfully to create the North Cascades National Park complex. In 1958 she and a few colleagues launched the Mount Rainier National Park Associates, the first volunteer watchdog group focusing on the mountain.

Dyer's commitment to protecting wilderness areas thrust her into the campaign for passage of the federal Wilderness Act, which became law in 1964. In fact, she suggested some of its language. Twenty years later she spearheaded the successful campaign to enact a wilderness act in Washington State.

Dyer knew the environmentalist and political leader Daniel Evans well. As governor of Washington, he had appointed her to a four-year term on the state's Forest Practices Board in 1974. While he was governor, she worked with him and her congressman to add the last seven miles of roadless coast, encompassing areas known as Point of the Arches and Shi Shi Beach, to the Olympic National Park. After Evans became a U.S. senator, Dyer says she persisted in her efforts to convince him "that before he stepped down from the Senate, he was the only person in Congress (then, earlier, or later) who could secure congressional wilderness designation for Olympic, Mount Rainier, and North Cascades National Parks—and he did it." Of all her achievements over the decades, Dyer cites these two as giving her the most personal satisfaction.

Polly Dyer has the reputation of being able to keep dialogue open during controversies, especially with the business sector. She claims patience, perseverance, and courtesy are the keys to being able to disagree without getting personal. Especially in the early days of her activism, she often found herself the sole woman and the only conservationist in a large meeting or on a panel. "I recall my naïveté when attending meetings about forest issues around 1952–53, often in a room of five hundred men, when it didn't occur to me that I was being conspicuous by asking questions from the floor."

Around that same time, she ended up (by accident, she says) on the governor's 1953–54 Olympic National Park Review Committee, appointed to decide whether a fourth of the park should revert to the U.S. Forest Service. "The committee had been 'stacked' with forest industry representatives [all men] until the Mountaineers and Olympic Park Associates met with and then wrote to the governor, pointing out that it needed a conservationist. Since the Mountaineers' president, who had then been appointed, couldn't participate, I as a fancy-free housewife took his place." The veteran environmental activist David Brower credits Dyer with swaying the review committee to keep the park intact.

Dyer does more than maintain a civil dialogue with the business community. In 1984 she helped found the Puget Soundkeeper Alliance to work with automotive shops, shipyards, and marinas to develop operating procedures that reduce the volume of pollutants going into the sound. This group also conducts a Soundkeeper training course for citizen volunteers, teaching them how to identify pollution cleanup needs. Dyer served as president of the alliance for two years and remains a member of its board.

Extensive dam building (more than fourteen on the Columbia River alone) left Washington's natural river ecosystems dysfunctional. Moreover, it all but decimated the state's once-thriving salmon industry, since the dams block fish from reaching their spawning areas. Removing dams where possible and restoring river basin ecosystems, especially those in the Olympic National Park, continue to be priorities on Dyer's list of conservation goals. This is not an easy venture, since it requires persevering over years to persuade the federal government to appropriate millions of dollars to purchase the dams from the private companies owning them. But Dyer says that "conservationists will continue to hang in there and work for appropriations to purchase the dams and hopefully see the Federal government recognize they have a major effect on a national park."

Considering the volume of Dyer's achievements as a volunteer activist, one might conclude she must have devoted all her time to these efforts. Not so. From 1974 until 1994 she had a full-time job as continuing environmental education director with the Institute of Environmental Studies at the University of Washington. As head of this outreach service, she developed background information on ecological issues—presenting both the pros and cons—and organized conferences for examining controversial questions. She also edited the institute's newsletter, *Environmental Outlook*. On top of all this, she pursued a college degree.

Dyer's formal education had been piecemeal. With a Coast Guard father assigned around the country, she attended ten different elementary schools. Lack of money precluded her attending college. Instead, she acquired secretarial skills. Not until she was forty-one did she undertake pursuing a college degree, part-time. Ten years later, after earning her bachelor's degree in geography (cum laude) from the University of Washington, she completed course work for a master's degree but did not finish writing her thesis. Now in her mid-seventies, Dyer says she "may yet tackle the subject, but not as an academic endeavor."

While she has never been physically threatened by opponents, intimidation sometimes occurs. Once at a hearing about whether to retain a section of Lake Quinault's northern shore within the Olympic National Park's southwestern boundary, "someone angrily challenged my right to have any say about the subject of protecting the area. In another instance I was 'set up' when agreeing to speak on a panel at the Pacific Logging Congress held in Victoria, B.C. I was the only environmentalist on the program and the only woman. The hosts seemed to be enjoying putting me on the spot."

What has kept Dyer going year after year in the face of constant opposition? She credits "a philosophy that if one gives up because of opposition or hostility or disappointments or setbacks, then one has lost—or is setting oneself up for losing. You just keep patiently persevering, coming back and starting over again, if that is what is required."

If she could undertake some issue campaigns over again, what would she do differently? Dyer cites her experience in proposing wilderness designation for Glacier Peak in one of Washington's national forests. She and her colleagues in the Mountaineers Club proposed boundaries she thought were reasonable and achievable. "The lesson: the U.S. Forest Service, then, and ever since, will consider a wilderness advocate's proposal the *maximum*, and reduce every wilderness area substantially. If I were to do it over: when it comes to wilderness, include *all* the possible de facto wilderness in proposals, since the opposition will seek to have the smallest, or no, wilderness whatsoever."

Although Dyer never had children of her own, she led a Girl Scout troop for a number of years, instilling an ecological conscience in the girls. Staying in touch with many of these women, she has observed that they later shared this ecological awareness with their spouses and children. She believes women need to take a more active role within their families and communities to effect the transformation necessary to save the planet from ecological ruin.

Over the past twenty-five years, more than a dozen regional and national conservation organizations, as well as academic institutions, have honored Dyer. And in 1994, when Governor Michael Lowry dedicated the new Olympic Coast National Marine Sanctuary, only one environmentalist shared the podium with him—Polly Dyer. Many people in Washington insist that the state's map would look much different today were it not for Dyer's years of volunteer work.

PAT WAAK (1943–)

The Population Issue and Environmentalism

In the mid-1960s, serving as a Peace Corps nursing instructor in Brazil, Pat Waak recalls

> sitting in the health clinic examining babies. A woman shuffled in who was bent and drawn; and I thought in my naïveté that this was a seventy-year-old woman bringing her grandbaby in for health care. Women never came in for health care themselves, they came to the clinic only when the babies were sick. As I interviewed her and examined the baby, I discovered that this was her baby, not someone else's, not her daughter's, that she was not seventy years old, she was thirty-five. She told me that in her reproductive life she had already had twenty-one pregnancies. That was the moment in which my life really shifted. And I became deeply committed to women's reproductive rights and the field of family planning. Sitting in that health post or going door-to-door talking to women, I came in contact with the stark reality that women had no say-so in when and how they got pregnant. A lot of the women practiced self-induced abortions.

The situations she encountered there, Waak says, demanded all the patience and perseverance she could muster.

Three decades later, on the cusp of a new millennium, is life for women in the developing countries better? In villages in northern India, for example, women can be located only by identifying the names of their father, brother, or husband. Obligated to serve their husbands, they view sex as another household chore and consider fertility as one of its health hazards. The rate of population growth in this area exceeds India's current overall annual growth rate of 1.9 percent.[13]

By contrast, in the state of Kerala in southwestern India, population growth has stabilized at zero, even though per capita incomes remain low. Leaders there carried out a plan tailored to the state's religious, social, political, and cultural characteristics. By supporting proper nutrition and health care, they lowered infant mortality; they made free birth control available; and they achieved a high rate of literacy, particularly among women.

The most successful family-planning programs involve women in both design and implementation. Pat Waak has seen this demonstrated in her work with women in Africa, Asia, Latin America, and the Middle East over thirty years, first as a Peace Corps volunteer and later through her service with the Office of Population in the U.S. Agency for International Development (USAID). While with the Peace Corps, in addition to teaching in a nursing school, she ventured into the Brazilian slums to give classes on child health to poor, uneducated mothers. For her, this is the reality of the population problem.

Demographers point out that although it took ten thousand generations for Earth's population to reach two billion in 1945, that number has grown to 5.5 billion in the two generations since World War II, and they predict another rough doubling in the next fifty years. Not everyone believes this doubling will occur.[14] Already the rate of population growth worldwide is beginning to decline. Unfortunately, part of the decrease is due to rising AIDS mortality and to reduced fertility caused by persistent organic chemical pollutants that disrupt hormone functions.[15] Nonetheless, it embarrasses Waak to note that currently the United States, a disproportionate consumer of natural resources, has the highest population growth rate of any major industrialized nation. At the same time, she believes humans are smart enough to avert an overpopulation disaster. She has dedicated her career to developing strategies and taking actions to help reverse the growth trend by improving the lot of women worldwide.

Waak served as an official U.S. delegate to the 1994 UN International Conference on Population and Development (ICPD) in Cairo. Earlier she had participated in the 1991 international women's conference in Miami, and she was one of the driving forces in the women's caucuses held before and during the Earth Summit in Brazil in 1992. At the Cairo conference, women were seen as people with the power to achieve stabilization in population growth—not as culprits of high fertility. After the Cairo conference, writing as chairwoman of the Population and Environment Working Group of the IUCN's Commission on Environmental Strategy and Planning,[16] Waak observed, "The deliberations and dialogue moved the issue of population away from purely demographic numbers to one with a human face. . . . The Cairo plan links reduced fertility and women's status, employment, and education—all factors which have proven to be related to family size."[17]

Nongovernmental organizations achieved an unprecedented level of recognition at the 1994 International Conference on Population and Development. In Waak's view,

> The United States made the first advance in increasing NGOs' role by including NGO representatives in the delegation at the second ICPD preparatory meeting in 1993. By the time that delegation reached Cairo, countries like the Philippines, Bangladesh, India, Egypt and China had added NGO representatives to their delegations as well. . . . The majority of the private sector delegates were women [Waak being one] with a variety of established careers—encompassing fields such as population, family planning, development, and the environment.[18]

She led a contingent of fourteen National Audubon Society staff members attending the parallel NGO forum in organizing a caucus of environmentalists from every region of the globe—the first such caucus since UN population conferences began convening two decades earlier.

In advance of the Cairo Conference, Waak and other designated NGO delegates had formed a coalition to organize public discussion meetings, generate publicity, testify at State Department hearings, review and propose changes to draft documents, and attend UN preparatory meetings. She says participating in this process was one of the most satisfying experiences of her career.

❧ As early as 1972 the National Audubon Society had dispatched a staff person, Frances Breed, to the first international conference on the environment in Stockholm. She lobbied unsuccessfully for including population concerns in the conference action program. Until her death in the mid-1980s, she continued to strive within Audubon to place the issue of population growth on the environmental movement's agenda.

In 1985, following three years with Columbia University's Center for Population and Family Health, Pat Waak joined the Audubon staff, picking up where Frances Breed had left off. Audubon—the first environmental organization to establish a department dedicated to the population issue—hired Waak as director to address population growth's impact on ecosystems and to connect women's roles and status with environmental concerns. She designed a five-year program focusing on public education and citizen mobilization.

Linking resource consumption and population growth, in 1987 she launched her education program by producing one of the first videotapes on population and the environment, *What Is the Limit?* She was also coauthor of a companion guidebook, *Where Do We Go from Here?* The video is still used in high schools and universities across the country. Other videos followed, including *Human Population and Wildlife: An Audubon Perspective* and *Finding the Balance*, the latter exploring how local communities can address the overconsumption and population issues. This early work led to her coproducing three hour-long television programs, which aired in the early 1990s on the Turner Broadcasting System. In 1994, to stimulate a dialogue with the religious community, she wrote a handbook called *Faith, Justice, and a Healthy World*, which articulated the importance of ethics and values to the issue of population and the environment.

Waak, who grew up in Texas as the oldest of four children, had decided to be a nurse when she was ten years old. She earned her nursing diploma in New Orleans in 1964. Since then she has augmented her schooling with courses in biology, public health, and politics at various universities. In 1997, she was awarded a master's degree in Jungian psychology. Concurrently, she raised two daughters born during her marriage to a fellow Peace Corps volunteer. Now a grandmother, and remarried to an Audubon colleague, she manages a staff of thirteen persons scattered over seven states and the District of Columbia. Her department's annual operating budget has grown to almost a million dollars. On the side, she lectures at universities, does consulting work, and serves on a number of task forces.

Waak regards the Audubon program Sharing the Earth as her most innovative project and one of her most rewarding undertakings. A skilled fund-raiser, she obtained a grant to carry out one of her visions—an international exchange program whereby managers from eight Audubon wildlife sanctuaries in the United States visited their counterparts in eight developing countries with high rates of population growth to study the effects of human activities on wildlife and habitat. Surprisingly, the study revealed that, notwithstanding its wealth and technology, the United

States had degraded its natural habitat more extensively than had many of the developing countries. The participating sanctuary managers' observations have been collected in the book *Sharing the Earth*.[19] The American Association for the Advancement of Science and the IUCN have subsequently used the project's results as a springboard for further studies.

Working on a controversial issue like population growth exposes Waak to occasional heckling during conferences and meetings. She has also received obscene phone calls. Even though she has maintained Audubon's neutral position on abortions—emphasizing women's health, education, and economic empowerment, along with their right to prevent unplanned pregnancies—she sometimes receives letters from persons who call her names and enclose ugly pictures.[20]

What helps her persevere when she encounters opposition and setbacks? Waak says she sustains herself through envisioning what the world *could* be like and through nurturing her connections with nature by spending time in unspoiled places. And she dwells on any small victories that come along. She is also encouraged by reminding herself that "we brought the whole issue to a new level. None of the environmental groups in 1985 were talking very much about consumption, population, and women's roles. Audubon made a major commitment to it. We opened the dialogue within the environmental movement."

She is heartened by the rapidly increasing success of NGOs around the world in influencing decision making. "It is from the NGOs that new ideas, approaches, and solutions are springing forth at the local, national, and regional levels."[21] These groups, she says, have a vision of what they want their communities to look like in the future and are addressing ecological issues. The effect of their actions ripples all the way to the national and international levels.[22] She also notes that in most of the countries where she has worked between 80 and 100 percent of the grassroots activists have been women. In her article for the *Colorado Journal of International Environmental Law and Policy* cited earlier, she names the Chipko and Green Belt movements in India and Kenya as but two examples.

Waak observes that rapid population growth, poverty, and environmental degradation feed on each other. Tackling these problems together holistically, she says, instead of attacking them separately, is the only way to climb out of our predicament. Like Ellen Swallow a hundred years ago (see chapter 2), Waak believes we need multidisciplinary education integrating the social sciences with the biological sciences in order to understand and confront the complexity of human beings' interactions with the globe's ecosystems. She anticipates that an increasing mass of citizens' groups whose members understand the consequences for our children of ongoing excessive consumption and overpopulation will trigger a reassessment and transformation of our values, lifestyles, and business practices. As she listens to college students remarking that our old economic system no longer works, she believes this transition has already begun.

In the late 1990s, although frustrated by declining support in Congress—which cut funding for worldwide family-planning programs by 86 percent for fiscal 1996—Waak nevertheless pursues what she believes to be her life's mission. The lack of commitment at the top led to her decision to work from the bottom up to influence decision makers. In 1997, using people power through her department's network of advocates and Audubon's base of over five hundred chapters, she launched the Population and Habitat Campaign. Its goal is to build a public mandate for family planning and to connect the issues surrounding population pressures with worldwide habitat and species loss and the consequent threat to human survival. Opinion polls show that three-quarters of Americans want the United States to do whatever is necessary to protect the environment. So by expanding the constituency of well-educated and well-trained grassroots advocates, building coalitions with groups outside the environmental movement, engaging the media, informing the people through public speaking, and maintaining a population policy specialist on Capitol Hill Waak believes attitudes in Congress can be brought around to conform to the commitments made at the Cairo conference in 1994. She designed the campaign to accommodate four levels of citizen involvement, ranging from highly trained, full-time, paid state leaders to interested citizen volunteers having limited time to contribute.

Leadership development and citizen activist training tools for the Population and Habitat Campaign include *What Is the Limit?* in an updated version; a new video, *Who's Counting;* a guidebook, *Population and Habitat for a New Millennium;* a three-part television series; extensive electronic communications, including a site on the World Wide Web; a special publication for middle school students; and a variety of newsletters and fact sheets. On 11 July 1997, World Population Awareness Day, members of Congress and their staffs had a chance to view *Who's Counting* at a screening on Capitol Hill.

When young people ask Waak what particular courses of instruction they should take to prepare them for environmental problem solving, she recommends first and foremost biology and environmental science, and then conflict resolution and consensus building, ethics, communications, and psychology. She excludes economics because the "traditional economics being taught in universities today really does not give any kind of flexibility to look for the alternative structures that will be essential for the next generation."

The real goddesses of Liberty . . . do not spend a large amount of time standing on pedestals in public places; they use their torches to startle the bats in political cellars.

ELLA S. STEWART
Nineteenth-century activist

Casting New Models

GRO HARLEM BRUNDTLAND (1939–)
Rises to Global Leadership

Norway's first woman prime minister, Gro Harlem Brundtland, was the first head of government anywhere to have moved into that role after serving as minister of the environment, also a world's first for a woman. How did she attain such heights of leadership?

Brundtland grew up in a family where dinner-table conversation centered on controversial public policy issues and where sexual equality was a given. Like her father—who had served in Parliament—she, too, became a physician, obtaining her medical degree from the University of Oslo.

As a pediatrician and feminist pioneer, with a master's degree in public health from Harvard, Brundtland first gained countrywide attention when, in the early 1970s, she assumed the role of national spokeswoman in the successful campaign to legalize abortion in Norway. Through speeches and articles, she recounted some of her patients' compelling experiences (concealing their names)—women she had treated while serving as medical officer for health services agencies in Oslo.

Recognizing her intellectual and political acumen, Norway's prime minister appointed Brundtland minister of health and social services in his cabinet. Shortly thereafter, in 1974, when she was only thirty-five years old, he reassigned her to the role of minister of the environment, where she served for four years.

In 1977, as an active member of the Labor Party, she won a seat in Parliament. During her years in the cabinet and her term in Parliament, her husband, a conservative newspaper editor, columnist, and scholar, managed their household and raised their four children, an unusual role reversal. Brundtland says hers is the first generation in which women can retain and express differences in background, political opinions, and social attitudes after they marry.

She resigned from Parliament in 1979 to concentrate on revitalizing the Labor Party and soon became its chairman. She instituted a model gender-balance policy, based on the principle of having a minimum of 40 percent female and 40 percent male candidates in every election. Then, in 1981, she served as prime minister for eight months, having been elected by the Labor Party to fill a vacancy.

Early on she felt as though everyone in the country expected her to fall on her face, but she hoped her radical course would make things easier for the next prime minister. Brundtland appointed women to almost half the cabinet posts, lengthened paid maternity leave to twenty-four weeks, and shortened the workweek to thirty-seven and a half hours. And as it turned out she *was* the next prime minister, being elected to a full term in 1986 and remaining the leading political figure in Norway for the next decade.

In 1983, between her terms as prime minister, while she was managing the day-to-day affairs of the Labor Party as its leader, the secretary-general of the United Nations asked Brundtland to establish and head an independent commission to address the global environmental crisis and to formulate an agenda for change. When he outlined what the General Assembly wanted the World Commission on Environment and Development to tackle, Brundtland thought the scope of the request was unrealistically ambitious. It did, however, reflect

a widespread feeling of frustration and inadequacy in the international community about our own ability to address the vital global issues and deal effectively with them. . . .

Since the answers to fundamental and serious concerns are not at hand, there is no alternative but to keep on trying to find them.

All this was on my mind when the Secretary-General presented me with an argument to which there was no convincing rebuttal: No other political leader had become Prime Minister with a background of several years of political struggle, nationally and internationally, as an environment minister. This gave some hope that the environment was not destined to remain a side issue in central, political decision making.

In the final analysis, I decided to accept the challenge. The challenge of facing the future, and of safeguarding the interests of coming generations. For it was abundantly clear: We needed a mandate for change.[1]

She and the commission's vice chairperson selected its other members—an interdisciplinary team from twenty-one countries, over half of them developing nations. The commissioners participated as individuals, not as representatives of their governments. Eight official "sponsoring countries" (not including the United States) covered the commission's expenses, with seven other nations and two foundations also making donations.

When the commissioners went to work, Brundtland later admitted, their disparate viewpoints seemed unbridgeable, especially as they related to population issues linked to the environment, to poverty, and to economic development. She notes, "We joined the Commission with different views and perspectives, different values and beliefs, and very different experiences and insights. After . . . three years of working together, travelling, listening and discussing, we present a unanimous report."[2]

The commission conducted public hearings on five continents, a trademark of its work. During its deliberations, which extended from 1984 to 1987, there were major environmental tragedies—Bhopal, Chernobyl, famine in the Horn of Africa, the explosion of a gasoline depot in Mexico City—all with heavy human casualties.

The commission's report, *Our Common Future*, defined "sustainable development" as organizing our affairs so as to meet our own needs without compromising the ability of coming generations to meet theirs. Consumption must remain within the bounds of what is possible ecologically. The report stressed that current practices are not sustainable—although they may show short-term profits, our children will bear the costs and may damn us, the commissioners wrote, for our profligate ways.

The commission projected that the transition to sustainable development would necessitate a massive shift in society's attitudes and objectives. Through education at all levels, institutional development, and law enforcement, individuals must be persuaded to act in the common interest. The blunt and sobering report—which soon became known as the Brundtland Report—forced the issue of global responsibility onto the international agenda. Its advice has been termed as important as a test-ban treaty in securing the earth's long-term welfare. In 1987 the British Broadcasting Corporation and PBS in the United States aired an eleven-part television series, *Only One Earth*, based on the report. It was an immediate best-seller as well, going through nine printings within scarcely more than a year. Whereas earlier initiatives had fallen short, *Our Common Future* made economic growth, debt, hunger, and poverty into environmental issues. It tied problems together and offered new models for integrated global solutions. The report urged greater citizen participation and greater democracy in international decision making, and more international cooperation in managing ecology and economics as interdependent factors.

Her role in shaping the report and introducing its conclusions to the media and to the public at international conferences thrust Gro Harlem Brundtland onto center stage. At forty-eight she became a worldwide environmental leader and began shuttling about the globe—speaking before the United Nations and appearing personally and on television in many countries. She advocated reversing ecological management practices from after-the-fact cleanup to up-front prevention. Futurists, experts in economic development and ecology, leaders in developing and developed countries—mostly men—recognized her as an influential force.

In 1989, Brundtland wrote in *Scientific American:*

> To secure our common future, we need a new international vision based on cooperation and a new international ethic based on the realization that the issues with which we wrestle are globally interconnected. This is not only a moral ethic but also a practical one—the only way we can pursue our own self interests on a small and closely knit planet. . . .
>
> Our generation is the first one to have seen planet earth from a dis-

tance. And from that perspective it is all too apparent that our species is dependent on a single tiny, fragile globe floating in space, a closed and vulnerable system.[3]

Addressing the United Nations in 1993, she pointed out that the world's poor "make minimal claims on our natural resources, while the more voracious North is consuming in a few decades what it has taken the planet billions of years to accumulate. This widening gap between the fortunate few and the powerless, impoverished majority is a destabilizing trend. It is both dangerous and morally unacceptable."[4] Brundtland noted that the average individual in North America consumes almost twenty times as much as a person in Asia (sixty times as much as a person in Bangladesh). She stressed that reducing consumption in industrialized nations does not have to mean lowering standards of living.

In her address, Brundtland also expressed disappointment that the momentum generated at the 1992 Earth Summit had been lost. In this regard, she mentioned the failure of the rich countries to follow through on their commitments to transfer additional financial resources to developing countries where environmental degradation is a survival issue. For example, between 1994 and 1995 the United States reduced its aid by $2.5 billion, sinking to fourth place among donor nations.[5] At the Earth Summit in 1992 most industrial countries pledged to strive toward increasing foreign aid from 0.35 to 0.7 percent of their gross national products. Instead, as of 1997, aid has slipped to below 0.3 percent.[6] Unless the North follows through with promised financial assistance, developing countries—preoccupied with the problem of feeding their people—will be hard put to find the resources and political will to put environmental safeguards in place.

In her 1993 UN speech Brundtland also remarked: "Experience shows that investing in women is one of the most cost-effective ways of promoting development. As mothers, as producers or suppliers of food, fuel and water, as traders and manufacturers, as political and community leaders, women are at the center of the process of change. . . . Women's education is the single most important path to a combination of higher productivity, lower infant mortality and lower fertility."[7]

Norway enjoys a worldwide reputation for its high level of environmental quality, especially clean air. Cabinet members, like many citizens, pedal their bicycles to work. And Brundtland sometimes skis from her Oslo home to her mountain cottage twenty-five miles away. Norway has few rich persons and few poor. Yet its foreign aid budget for developing countries is one of the highest per capita among industrial nations.

In October 1996 Brundtland—now a grandmother—announced that after more than a decade as prime minister she was stepping down. She leaves the legacy of having pointed the way toward a planet-healing political context, a sustainable development model.

HAZEL HENDERSON (1933–)
Futurist

In 1975, when Hazel Henderson appeared before the Joint Economic Committee of Congress to criticize traditional economics and present her views on the economic transition she foresaw, committee members laughed, protesting that they could see no evidence of such a transition. In 1991, however, there was no laughing when Henderson appeared "as an expert" at a London meeting of thirty-five former heads of state to discuss "Economics in Transformation: Limitations and Potential of the Transition Process." Nor was there laughter in 1995 when, at Mikhail Gorbachev's invitation, she convened Roundtables on Twenty-First-Century Economics for his first annual State of the World Forum in San Francisco.

One of this century's foremost "high-risk social innovators," Henderson says she likes to "intercept policy debates anywhere in the world . . . offering new directions, expanded contexts, connections and possibilities for creating win-win solutions in our ever-shrinking, more crowded, polluted planet."[8] In the 1970s, though, her multidisciplinary approach was less than appreciated. Her articles in business journals during that decade (including "Ecologists vs. Economists" in the *Harvard Business Review*) led to her being labeled the "most dangerous woman in America" by corporate CEOs. Economists called her the lady who was poisoning the well. Since then, Henderson has become a respected and sought-after authority on global economics and human development issues. She has written four books—translated into eight languages—and over two hundred and fifty articles. Her syndicated column appears in twenty-seven languages in four hundred newspapers worldwide—though not in any U.S. paper. The *Utne Reader* includes her in its list of one hundred and twenty visionaries who could change our lives.

More than twenty-five years ago, Henderson embarked on a crusade to persuade countries to overhaul the way they each measure economic growth and human development. Calling herself an anti-economist, she challenged the traditional economic assumptions—based on obsolete nineteenth-century industrialism—that govern gross national product (GNP) and gross domestic product (GDP) statistics. For example, the GNP overlooks half the production going on in the world—unpaid work, contributed mostly by women; as Henderson notes, they "produce half the world's food and manage seventy percent of its small businesses, but receive only ten percent of the world's wages and own only one percent of the world's property."[9] Nor does the GNP accounting system classify natural resources as assets; for that reason, it fails to deduct from growth statistics the cost of diminishing or destroying natural resources during industrial processes.[10] Ironically, the system treats cleaning up pollution and repairing damage (such as the *Exxon Valdez* oil spill and the Chernobyl nuclear accident) as additional production.

Henderson believes economics is merely politics in disguise. "We need to send the economists back to school to learn more sociology, cultural anthropology and ecology before we trust their calculations" (*PP*, 222). They have been badgering environmentalists for decades to learn economics, insisting that only "experts" can understand how GNP and GDP indexes are constructed. Henderson and other environmentalists suggest, on the contrary, that it is the narrow focus of economics courses that needs to be broadened to include training in the environmental and social sciences.

After emigrating from the United Kingdom in 1956, Henderson lived for a time in New York City. The first Clean Air Act (passed by Congress in 1955) was so weak it had done little to clean up cities. Henderson says she had to give her daughter baths after she played in the park just to get the soot off her. The dense smog she observed from her apartment window was also alarming; it reminded her of London's 1952 smog crisis, during which four thousand people died.

In 1964 Henderson began a letter-writing campaign. Her first correspondent, Mayor Robert Wagner, maintained that what she saw was not pollution but mist rolling in from the sea. Undeterred, she wrote to the presidents of the major broadcasting networks—sending copies to New York City's senior elected officials and to the chairman of the Federal Communications Commission (FCC)—urging them to include an index of air pollution in the weather reports each day. She reminded them that the Communications Act of 1934 required that stations broadcast information in the people's interest. Chairman Newton Minow of the FCC replied enthusiastically, saying he would be interested in how the networks responded to her suggestion. Henderson sent copies of Minow's letter to the networks' presidents.

About five weeks later, to her surprise, a vice president from ABC in New York called her, saying he thought the index was a great idea. Henderson pointed him in the direction of the New York City pollution control commissioner. Within six weeks, WABC-TV in New York City began broadcasting an air pollution index, with the CBS and NBC affiliates, the *New York Times* and other newspapers, and radio stations quickly following. Before long, reporting air pollution information became a standard part of weather broadcasts nationwide.

As a follow-up, Henderson and two friends organized Citizens for Clean Air to warn New York City residents about air pollution's health hazards. With the pro bono help of a fledgling advertising agency, Citizens for Clean Air launched an educational campaign. The group's membership swelled from fewer than one hundred to twenty-five thousand persons—three-fourths of whom were women.

Henderson had dropped out of school at age sixteen. After coming to the United States, wanting to understand the culture of her adopted country, she began educating herself. She studied sociology, ecology, anthropology, general systems theory, history, and philosophy. Assuming herself to be unemployable, she concluded she would have to invent her own job. And so Henderson decided to teach herself economics. Following extensive reading (including college economics textbooks), she began to write articles. In addition to the *Harvard Business Review*, she published pieces in the *Financial Analysts' Journal* and the *Columbia Journal of World Business*. She exposed flaws in traditional economic theory, including its definition of real wealth and its methods of calculating growth.

In the late 1970s she realized with alarm that economics as taught and practiced was, as a policy tool for citizens' groups, beyond redemption. "I saw economics lead its practitioners and citizens alike into a form of brain-damaging indoctrination. Horrified, I pulled back from activism" (*BW*, 144). After fifteen years of trying, she gave up on reforming the system.

She returned to research, studying the evolution of human ethics, altruistic behavior, chaos theory, conflict resolution, "cultural DNA codes," cooperation, and win-win solutions. Not tied financially to any business or government, she remained independent—as did her conclusions. Her research led to her first two books—*Creating Alternative Futures: The End of Economics* (1978) and *The Politics of the Solar Age: Alternatives to Economics* (1981; rev. 1988). Of the latter, she observes, "My analysis was vilified by economists as wrong-headed and absurd. I learned to interpret this as evidence that I was hitting home" (*BW*, 144).

In 1974 Congress had created the Office of Technology Assessment (OTA) as one of its research arms. The mission of the office was to assess the social, environmental, economic, and political impacts of emerging technologies before they were fully developed. Henderson conducted a successful campaign to be appointed to the OTA's advisory council and served as the only citizens' group advocate. "My 'Ph.D.' course," she reflects, "was six years on the Advisory Council of the U.S. Office of Technology Assessment and on panels of the National Science Foundation and the National Academy of Engineering from 1974 to 1980" (*BW*, 76). In a budget-cutting frenzy in 1995, the 104th Congress eliminated the OTA.

In 1975, recognizing her talent for taking the long view, Jimmy Carter appointed her to his election task force on economics. She recommended expanding the President's Council of Economic Advisors into a multidisciplinary Council of Social Science Advisors. Ronald Reagan's election coincided with the publication of *The Politics of the Solar Age*. When, as president, he canceled solar energy tax credits, women and men starting companies based on solar energy, energy efficiency, recycling, and biomass "realized there wasn't going to be any politics of the Solar Age, but rather only politics of denial. . . . People began to write to me and load me up with business plans. I got worried that there was no place to take those business plans, however good they were. So in 1982 I got involved in the socially responsible investment movement" (*PP*, 196). The Calvert Social Investment Fund appointed Henderson to its advisory council, on which she has served ever since. She came to know young solar-age activist entrepreneurs thinking through their new businesses. Socially responsible investment funds have been growing faster in size than any other market segment. In a 1997 survey, the Calvert Group determined that 81 percent of Americans would be more inclined to invest in a company if they knew it to be environmentally responsible.

In 1970 Henderson became active on the board of the Council on Economic Priorities (CEP); at that time, it was considered radical by corporate and political leaders. Now well respected as a nonprofit public interest research group, CEP monitors corporate performance and ranks companies in eleven categories of social responsibility. Its booklet *Shopping for a Better World*, updated each year, has sold over seven hundred thousand copies. More recently she became a fellow of the World Business Academy, a group made up mostly of international "CEOs who have had a change of heart and realize how important socially responsible operation of business is for all of their stakeholders, not just their stockholders."

In Henderson's view, we are witnessing a revolution in industrial redesign and a reinvention of management. As an example of what can happen when industries pool their efforts, she points to Kalundborg, Denmark, "which thinks big and in systems terms. Here, industrial wastes and waste-process heat are exchanged between a power plant, an oil refinery, a pharmaceutical manufacturer, a plaster-board factory, a cement producer, farmers and the district supplier of home heating. These win-win-win approaches show the future potential of organizing common systems through cooperation—to complement the competitive market place. This paradigm shift is occurring rapidly" (*BW*, 44).

Thus, a new ethical capitalism has been developing in mature industrial countries. Henderson sees the competitive win-lose paradigm of market economies being replaced by a win-win combination of competition and cooperation governed by ethics, trust, and creativity. "Competing by cooperating and standard-setting will hallmark twenty-first century economies" (*BW*, 275). She points out that "trickle-down GNP growth policies are failing while trickle-up micro-enterprise-based, rural development with locally controlled technologies and resources are succeeding" (*PP*, 126).

Henderson believes a primary cause of the world's social, political, ecological, and economic problems is that technological advances have outpaced social innovation. For example, now that global trading in currencies and bonds is carried on through computers, she observes that $1.3 trillion swirls around the planet each day in what she refers to as a global casino. This results in global financial markets driving the policies of nation-states. Yet capital markets do not account for social and environmental costs. "If the global casino's stranglehold on local communities continues, local information currencies, regional payment unions, virtual banks and E-cash on the Internet, global barter and countertrade, already estimated at twenty-five percent of all world trade, may simply break the global money cartel by end-running it" (*BW*, 55).

Henderson proposes that international agreements and regulations be established to control these transactions. Even though she criticizes the United Nations for being sexist, she believes that, with some reform, the UN is the one entity already in place "and truly positioned to convene, foster and broker all the actors and institutions in government, business, finance, academia and the global civil society" (*BW*, 54). As a founding member of the Commission to Fund the United Nations, she coedited and wrote the introduction and a paper for the commission's first report, issued in 1995. Entitled *The United Nations: Policy and Financing Alternatives,* the report sets forth innovative models developed by visionary leaders of how the trade and capital markets of the world can become more efficient, more orderly and democratic, and more environmentally and socially responsible.

Henderson condemns military spending and the international arms trade. She notes that the five main arms-dealing countries are the five permanent members of the UN Security Council—with the United States leading in arms sales. She contends that the annual amount necessary to solve the major global problems relating to human need and the environment adds up to about one-fourth of the world's annual military expenditures (*PP*, 230–33).

A dominant theme in her writings and speeches during the past thirty years has been the need to reformulate how the GNP is calculated. It is, she says, a weak indicator for quality of life. "One of the most critical errors of economic theory has been the omission of the informal, unpaid sectors from its model (parenting, home management, do-it-yourself home-building and repairs, mutual aid, volunteering, food raising, caring for the sick and elderly, bartering, etc.) . . . the unseen half" (*PP*, 101). In its 1995 report, the United Nations Development Programme (UNDP) estimated the value of this unseen work at sixteen trillion dollars, eleven trillion of it produced by women.[11] This sort of productivity is not recognized in any national accounting model. Furthermore, different cultures and their value systems do not all fit into one "correct" model for calculating economic growth and development.

To augment GNP statistics, new social indicators are emerging for gauging human development; these new scorecards reflect rapid global changes. The demand for such new indicators was popularized at the 1992 Earth Summit when some twenty-six thousand NGO representatives and activists adopted sustainable development as their rallying cry. Henderson herself, using an interdisciplinary process and in collaboration with the Calvert Group, Inc., has adapted her Country Futures Indicators to form the Calvert-Henderson Quality of Life Indicators for the U.S.A. as an alternative to the gross national product. Her model reformulates the GNP to correct errors and provide more information, and it establishes complementary indicators of progress toward society's goals (including such things as literacy levels, health and nutrition, safety from crime, environmental quality, biodiversity, and the status of women and minorities).

Henderson cites Costa Rica as a country demonstrating leadership in defining new indicators for sustainable development. Poor by GNP standards, Costa Ricans take pride in emphasizing their social values—health, education, biological diversity, justice, and peace. The country's literacy rate is higher than that of the United States. Costa Rica abolished its military in 1949—the first country to do so.

In 1990 the United Nations Development Programme began issuing an annual Human Development Index, ranking 173 countries in various aspects of social progress, such as life expectancy, literacy, and purchasing power. Henderson acts as an editorial advisor for the *World Paper*, a monthly insert in twenty-eight influential newspapers in five languages that covers new indicators being developed.

Even the holdout World Bank has begun to turn around. In 1995 it released its new Wealth Accounting System. This index reckons four kinds of assets as constituting the real wealth of countries: natural environmental resources, produced assets (infrastructure and financial assets), human resources (healthy, educated people), and social capital (families, institutions, and communities).

Henderson views the world's current social and economic chaos as evidence of a paradigm shift attendant upon humans' maturing as a species. Breakdowns and crises inevitably herald breakthroughs, and stress is evolution's tool, she believes, to make humans grow up. But, she says, intellectual lag pervades all our institutions—academia, government, business, labor, religion. It takes at least ten years, Henderson observes, before politicians react to social change. "Most of our long-held beliefs about money, wealth, productivity and efficiency, and our notions of progress are rooted in immature, often infantile states of mind—easily manipulated by politicians and advertisers. . . . Luckily, individuals learn faster than institutions and many people may already be ahead of their leaders" (*BW*, 153).

In her third and fourth books, *Paradigms in Progress* and *Building a Win-Win World*, she discusses this evolutionary process. Henderson sees the passage out of industrialism and patriarchy, out of the solar age, past the information age and into the Age of Light as offering undreamed of opportunities for creativity and innovation (fiber optics, lasers, optical scanning, optical computing, photovoltaics, and so on), while some industrial sectors become obsolete.[12] "The newest enterprises must address environmental restoration and enhancing where possible the performance of eco-systems. . . . Bioremediation and desert-greening will be big business in the twenty-first century, a point I have been emphasizing since the 1970s" (*PP*, 270).

Henderson credits global grassroots activism with accelerating the evolutionary process toward sustainable development, abetted by the Internet. Business and government leaders who once spurned this powerful constituency (now numbering millions of groups)—considering them amateurs and troublemakers—today concede they cannot lead or govern without consulting these effective problem solvers. There are now concurrent NGO summits with every gathering of international leaders.

The proliferation of business codes of conduct in the 1990s also signals the beginning of change in the corporate community. Henderson notes that in 1971, when she wrote her first article for the *Harvard Business Review*, courses in ethics did not exist in business schools. Now, such courses are not only common but frequently required.

Henderson travels the world, lecturing at seminars and talking with political and business leaders. For example, in the course of just one month in 1996 she participated in a conference, Business and Society, in Hyderabad, India, sponsored by the World Business Academy, the International Labor Organization (ILO), and local companies. Then she and other leaders of the Social Funder Network in Europe met with President Václav Havel of the Czech Republic; she found that he "really sees the need now that we are past the capitalism versus communism cold war to be discriminating about what kind of capitalism we embrace, something which I have been promoting all along—small entrepreneurial capitalism with employee ownership plans and management by all their stakeholders, as I wrote in *Creating Alternative Futures*." In addition to writing and lecturing, Henderson teaches a course in ecological economics at Schumacher College in the United Kingdom. She also participates as an advisor or board member in many organizations, including the Worldwatch Institute, the Americans Talk Issues Foundation (an independent polling group), and the World Future Society.

Although Henderson concedes that her writings and lectures have not had much effect on mainstream economics in academia, her influence on socially responsible business leaders has been outstanding. Heads of new businesses tell her they grew up on her first book, *Creating Alternative Futures*. Ironically, that book received little recognition, one reason being that potential reviewers told her that, even though they wanted to write about the book, they could not because she had admitted in the first paragraph that she had not gone to college. Otherwise, her biggest obstacle has been trying to obtain coverage in the mainstream print and broadcast media. She says this has frustrated her a great deal.

What keeps Henderson going? "I trust my perceptions and my analytical ability. I feel really in touch with the planetary and political processes. Also, I have a passion to communicate. I HAVE to communicate what I see and hear and predict or else I would just die. So my life is about communicating my message. That's what really keeps me going. Writing books has given me my greatest satisfaction in life. Through them I have identified my community and my intellectual family all over the world. My books were lifesavers and great delights because of the people I met through them."

"Environmentalism and its underlying eco-philosophy," Henderson declares, "constitute a major overarching paradigm that will from now on compete with economism and industrialism in both East and West, as well as North and South, in the twenty-first century. . . . The hour is late and the stakes higher than ever, but we have no other course than to learn how to intervene more responsibly to heal the Earth" (*PP*, 75, 57).

HELENA NORBERG-HODGE (1946–)
Development plus Preservation

When it comes to cultural progress and economic development, does one size really fit all? Should the Western-style monied economic and technological monoculture, driven by dissatisfaction and competition, be imposed everywhere, ignoring regional diversity and local conditions and needs? Does the Western brand of development too easily allow exploitation by multinational corporations? Must development mean destruction of traditional cultures? Might there be a different path into the future for developing countries?

The Swedish-born linguist and anthropologist Helena Norberg-Hodge explores these questions in her book, *Ancient Futures*, which has been translated into twenty-five languages.[13] In 1975 she traveled to Ladakh to study the language and culture. Living with Ladakhi families and participating in their daily work, she mastered the language within the first year. The terms "ecology" and "sustainability" meant little to her then, but (observing what Western development meant for Ladakh) she soon became a full-time environmental activist.

In the early 1970s, Ladakh had barely begun to be touched by Western influence. Situated on the Tibetan plateau on the north side of the Himalayas, its inaccessibility, lack of resources, and harsh desert climate had shielded the region from colonialism and development. Ladakhis lived as they had for over a thousand years in a nonmonied subsistence economy. They had attained almost complete self-sufficiency in spite of scant resources, importing only salt, tea, and a few metals for tools and cooking utensils. They had figured out how to grow barley at twelve thousand feet during only a four-month growing season and how to raise sheep, goats, and yaks at high altitudes. They produced their own vegetables, spun and wove wool for clothing, made warm and sturdy footwear from yak skin, and sun-dried dung for fuel. They practiced healing with natural plant remedies. They built their homes from mud bricks. Lacking significant rainfall, they channeled glacial snowmelt to irrigate their crops. Repeated recycling of everything precluded waste disposal problems.

> Where we would consider something completely worn out, exhausted of all possible worth, and would throw it away, Ladakhis will find some further use for it. Nothing is ever just discarded. . . . Ladakhis patch their homespun robes until they can be patched no more. . . . When no amount of stitching can sustain a worn-out robe, it is packed with mud into a weak part of an irrigation channel to help prevent leakage. (*AF*, 25)

Ladakhis—both men and women—learn a broad range of skills to meet their needs. With few exceptions, none is a specialist. In a cooperative community, they work together, bartering and sharing. Often singing as they work, they entertain themselves making music, dancing, telling stories, doing theater, and participating in festivals.

> Their sense of joy seems so firmly anchored within them that circumstances cannot shake it loose. . . . As I return each year to the industrialized world, the contrast becomes more and more obvious. . . . The Ladakhis . . . seem to possess an extended, inclusive sense of self. They do not, as we do, retreat behind boundaries of fear and self-protection; in fact

they seem to be totally lacking in what we would call pride. This doesn't mean a lack of self-respect. On the contrary, their self-respect is so deep rooted as to be unquestioned. . . . I have never met people who seem so healthy emotionally, so secure, as the Ladakhis. The reasons are, of course, complex and spring from a whole way of life and world view. But I am sure that the most important factor is the sense that you are part of something much larger than yourself, that you are inextricably connected to others and to your surroundings. (*AF*, 83–85)

When India opened Ladakh to tourism and modernization in 1974, rapid changes ensued. Norberg-Hodge compares the abrupt onslaught of tourists to an invasion of aliens from another planet. Before they first encountered tourists, Western films, and radios, she remarks, the people of the ancient Ladakhi culture never considered themselves poor. True, their gross national product measured zero and they had almost no money. They needed little. But seeing Western tourists in their modern clothes spending as much as a hundred dollars in a single day (the amount a Ladakhi might spend in a year), began to erode the self-image of Ladakhis, especially that of impressionable teenaged boys. Designer sunglasses, tight jeans, factory-made shoes, radios, cigarettes, cars, and living in concrete houses became status symbols. Possessing these things appeared to them the way "to be somebody." They began to be ashamed of their own culture. Some abandoned their rural community, seeking jobs in Leh, the capital.

In Ladakh I have known a society in which there is neither waste nor pollution, a society in which crime is virtually nonexistent, communities are healthy and strong, and a teenage boy is never embarrassed to be gentle and affectionate with his mother or grandmother. As that society begins to break down under the pressures of modernization, the lessons are of relevance far beyond Ladakh itself. . . .

It may seem absurd that a "primitive" culture on the Tibetan Plateau could have anything to teach our industrial society. Yet we need a baseline from which to better understand our own complex culture. In Ladakh I have seen progress divide people from the earth, from one another and ultimately from themselves. I have seen happy people lose their serenity when they started living according to our norms. As a result, I have had to conclude that culture plays a far more fundamental role in shaping the individual than I had previously thought. (*AF*, 4, 5)

With tourists came road construction, hotels, and schools. The schools—imitations of an Indian leftover of the British educational system—use textbooks written by people who have never been to Ladakh, and Norberg-Hodge adds that there is nothing Ladakhi about the schools. What the children learn is irrelevant to their daily experience and most of the material is never used once they leave school.

She concedes that some aspects of the Ladakhi culture need to change: literacy rates heightened, infant mortality reduced, life expectancy extended, and communication with the outside world widened. And enhanced creature comforts like better heating during the severe winter would be welcome. But the development being pressed on Ladakh, she says, produces serious environmental problems that, unless checked, will lead to irreversible degradation. For example, the government now encourages farmers to switch to monoculture and grow cash crops for export rather than plants for local subsistence. Norberg-Hodge contends this is a mistake. "The spread of the industrial monoculture is a tragedy of many dimensions. With the destruction of each culture, we are erasing centuries of accumulated knowledge, and as diverse ethnic groups feel their identity threatened, conflict and social breakdown almost inevitably follow." Furthermore, even though Ladakh has almost no pests, farmers—unaware of attendant hazards—are pressured to use BHC, a pesticide more toxic than DDT.

> To Western eyes, Ladakhis look poor. Tourists can only see the material side of the culture—worn out woolen robes, the *dzo* [a cow-yak cross] pulling a plough, the barren land. They cannot see peace of mind or the quality of family and community relations. They cannot see the psychological, social and spiritual wealth of the Ladakhis. . . . Every day I saw people from two cultures, a world apart, looking at each other and seeing superficial, one-dimensional images. Tourists see people carrying loads on their backs and walking long distances over high mountain passes and say, "How terrible; what a life of drudgery." They forget that they have traveled thousands of miles and spent thousands of dollars for the pleasure of walking through the same mountains with heavy backpacks. They also forget how much their bodies suffer from lack of use at home. . . . Some will even drive to a health club—across a polluted city in rush hour—to sit in a basement, pedaling a bicycle that does not go anywhere. And they actually pay for the privilege. (*AF*, 95, 96)

Since 1975 Norberg-Hodge has lived part of each year in Ladakh. She says watching the modernization of Ladakh made her see her own culture differently. She found herself constantly being asked about life in the West. So, as a Westerner who could speak the language, she undertook an educational program to present a more accurate picture of Western modernization than the impression gained by watching tourists and movies. One example is a play she wrote in collaboration with a Ladakhi. *Ladakh: Look before You Leap* tells the story of a young Ladakhi who rejects his culture and tries to live like a Westerner. When his grandfather becomes ill, he brings in an American-trained Ladakhi physician. Plying the doctor with questions about life in the United States, the young man listens as the doctor tells him:

> In America the most modern people eat something they call stone-ground whole meal bread. It's just like our traditional bread, but there it's much more expensive than white bread. People over there are building their houses out of natural materials, just like ours. It's usually the poor who live in concrete houses. And the trend is to dress in clothes with labels saying "100 percent natural" and "pure wool." The poor people wear polyester clothes. It's not what I expected at all. So much that is modern in America is similar to traditional Ladakh. In fact, people used to tell me, "You're so lucky to have been born a Ladakhi."

Norberg-Hodge notes that "[f]ive hundred Ladakhis crowded the auditorium in Leh for the play's premiere, which was a great success. Afterwards, local leaders, including . . . the development commissioner and highest-ranking official in the local administration, made speeches about the importance of cultural self-respect" (*AF*, 170–71).

When Norberg-Hodge told Ladakhis that Westerners sometimes had to go to a doctor (psychiatrist) because they were so unhappy, their mouths dropped open in astonishment.

In addition to her educational program for the Ladakhis about all aspects of the Western lifestyle, Norberg-Hodge produced a videotape about Ladakh's people and culture and the impacts of development. It was her hope to share it not only with people in Northern countries but with those in Southern countries considering alternative models for development. So far, the audio portion has been translated into twenty-five languages, and the videotape has been viewed by millions worldwide. She has tried to reach a larger audience in other ways as well.

> By 1980 my activities in both Ladakh and the West had grown into a small international organization called the Ladakh Project,[14] which in 1991 became the International Society for Ecology and Culture [based in the United Kingdom]. We seek to encourage revisioning of progress toward more ecological and community-based ways of living. We stress the urgent need to counter political and economic centralization, while encouraging a truly international perspective through increased cultural exchange. (*AF*, 171)

In 1983 Norberg-Hodge organized a group of concerned Ladakhi thinkers interested in exploring a more sustainable development model. She and this group, the Ladakh Ecological Development Group (LEDeG), engage in "counter-development" activities to stem the tide of greater specialization, centralization of production, and control from and decision making in faraway places. Working closely with members of the broader community, they have introduced Ladakhis to appropriate technology. When Ladakhis began buying government-subsidized imported coal and wood for home heating at prices they could ill afford, Norberg-Hodge obtained permission from the Ladakh planning commission to undertake a pilot project demonstrating a passive solar energy home-heating system. Mud-brick dwellings lend themselves well to absorbing and storing solar energy. The system involves painting one south-facing exterior wall black and installing a double glass wall against it on the outside. Vents along the top and bottom of the adobe wall carry the heat by convection into the home's interior.

Norberg-Hodge also introduced solar ovens, solar water heaters, and greenhouses, the latter enabling people to grow vegetables during the long winter. With increasing interest in electricity, LEDeG's technical program has focused on developing microhydro installations for home lighting. The villagers the program serves participate in all the projects— for example, choosing sites for turbines, building a holding tank, and attending workshops on how to operate and maintain the installation. As Norberg-Hodge points out,

All these technological alternatives make sense economically, environmentally, and culturally. By encouraging a more human-scale and decentralized development pattern, they actively support traditional cultures rather than destroying them. And they are not "technologies for the poor," only suited to the underprivileged. As we do our best to make clear, nonpolluting appropriate technologies based on renewable energy are *not* something second-rate, but highly effective and efficient solutions to the long-term needs of both developed and developing countries. (*AF*, 173)

LEDeG produced the first book in the Ladakhi language on the subject of ecology. Headquartered in Leh in the Center for Ecological Development (which Norberg-Hodge established in collaboration with Ladakhis), LEDeG as of 1992 had a staff of forty and had become the most effective and influential nongovernmental organization in Ladakh. Consecrated by His Holiness the Dalai Lama in 1984, the center itself exemplifies the use of alternative technologies with solar heating, a wind-powered generator, solar hot water, a greenhouse, and a workroom where appropriate technologies are produced. The center's restaurant serves food cooked in its solar ovens.

The Center for Ecological Development conducts wide-ranging educational programs on eco-development through seminars, training workshops, radio broadcasts, films, and publications. A handcrafts program begun in 1989 encourages local self-reliance by offering an outlet for items produced at home during the winter. Without abandoning village life, people can earn money. Tourists, now numbering over fifteen thousand each summer, can buy souvenirs made by Ladakhis, and they can watch wood carvers, weavers, silversmiths, tailors, and embroiderers at work.

Because Ladakh lies in a strategic border zone, foreigners normally are not allowed to work or live there. An article in the *Hindustani Times* during Norberg-Hodge's second year in Ladakh described her as a mysterious woman who "has picked up the language in a suspiciously short time," imputing some sinister nature to her presence. "Despite continued high-level government backing, including personal meetings with and letters of support from prime ministers and state governors, several intelligence officers remained convinced I was a C.I.A. agent collecting information about this sensitive border area" (*AF*, 176).

The village families with whom Norberg-Hodge lives during her periods in Ladakh each year are Buddhists, although about half the population of Leh is Muslim. Traditionally, and until development began, the two groups lived in harmony, even intermarrying. But contention arose in 1989, leading to fighting in Leh; police shot four persons and afterward imposed a curfew. The authorities accused her of causing the rioting, and she also received threats on her life. On another occasion Norberg-Hodge was arrested. "I was very frightened. It was in a remote part of Ladakh and the police behaved badly." Once she had to flee Ladakh altogether.

What can be done to alter the course of development in the world? Norberg-Hodge advocates launching a global re-education campaign to counter the distorted images of the industrial model being promulgated by multinational corporations and by Western governments. She believes individuals need to be able to make fully informed choices about the kind of future they want. In a chapter of her book, "Counter Development," she outlines a course of action that might check the rush toward unsustainable development worldwide. She remarks, for example, "One of the most effective ways of turning destructive development into genuine aid would be to lobby for widespread support and subsidies for decentralized applications of renewable energy" (*AF*, 164). She also points out that

> [f]arming provides the most basic of all human needs and is the direct source of livelihood for the majority of the people in the Third World. Yet the status of the farmer has never been lower. . . . If present trends continue, the small farmer may well be extinct in another generation. It is imperative that we reverse these trends by giving agriculture the prominence it deserves and actively seeking to raise the status of farming as an occupation. . . . Small farmers would be better off if emphasis were placed on food production for local consumption, rather than on crops for export. . . . They would also benefit if support were shifted away from the use of pesticides and chemical fertilizers to more ecologically sound methods. (*AF*, 165)

Norberg-Hodge maintains that people from developing countries who have lived in the West for a time and then returned home are the most credible witnesses to what Western modernization does to a culture socially and environmentally. They "are the ones who are most convinced of the need to find another development path and are the most interested in our work."

In assessing the current level of balance in Ladakh between maintaining the traditional culture and using appropriate technology to make life less harsh, Norberg-Hodge remarks,

> As of the summer of 1996, I would say that the situation in Ladakh is quite hopeful. After twenty years of exposure to development, many educated and modernized Ladakhis have come to realize that there are a number of drawbacks and problems associated with modernization, and there has been a tremendous increase in interest in our work from this group of people. . . . Not only is interest growing dramatically in Ladakh but the leaders of the Ecology Group [LEDeG] that I helped to start are now actually heading a new, semi-independent, local government. That means that the policies of the Ecology Group have become the policies of the Ladakhi government, making it a very exciting time in Ladakh.

She senses the emergence of a complementary trend in the West—a trend that has been growing for some time—toward a less consumptive but more satisfying lifestyle. Individuals are once again getting in touch with their deeper human nature. She is also pleased to see greater questioning of corporations' influence on systems of governance,[15] along with criticism of the antidemocratic, usually secret way in which current free trade agreements have been formulated. She believes it is vital for nations to renegotiate these international treaties to eliminate corporate domination, in which a few persons have effective control over enormous amounts of wealth.

Norberg-Hodge says that the most gratifying aspect of her work has been the interest among non-Western cultures (where centralized energy-generation systems have not yet established a stranglehold) in renewable energy systems. She has also received great satisfaction from how women respond to her book.

> Women seem to maintain a deeper spiritual connection to the living world, to the living fabric of human society as well as nature. The anti-life forces of technology and economic growth are more alien to them, they

don't identify with them as much as men in industrial society seem to, and therefore their commitment to changing the direction of industrial society seems to be considerably greater. . . . Women are more at ease with perceiving reality as a continuum and as a process, and because they do not rigidly separate mind from heart, emotion and bodily experience, they tend to perceive the world around them in a more holistic way. Men on the other hand have developed a so-called analytical world-view which is far more fragmented. . . . I find again and again that women quite intuitively reject the blind dogma and religious fervor around technological development and economic growth, but they often are not part of the more articulate opposition to it. They often feel weakened and demoralized by being told that they don't understand, so they should just keep quiet and accept what's happening. Part of our work is trying to involve more women in looking at the broader contours and principles of the global economic system in order to be more articulate and convincing in their opposition to it.

Norberg-Hodge, who grew up in Sweden in an affluent family, now maintains her home base in the United Kingdom. Although her husband trained to be a barrister, for the past fifteen years he has also worked with her on the Ladakh project.

What keeps her going when she encounters opposition or setbacks? "My deep faith in human nature. My conviction that children are born essentially seeking love, rather than hostility and competition. I believe the desire for cooperative and loving relationships is deeper than any desire for the opposite."

Does she entertain hope that humans will be able to save the planet from ecological ruin?

I don't think we have much time, but I still have hope that we can halt the mad suicidal race we are engaged in. I have become convinced that the engines of destruction are essentially the financial markets and transnational corporations that drive the economy and are increasingly shaping politics. . . .

But in the last few years, awareness has been growing rapidly, and I'm very excited about an organization that I'm a founding member of called the International Forum on Globalization,[16] which consists of about sixty thinkers, activists, and writers from around the world. We have been working as a group to alert people to these issues at an international level, while also doing public awareness work in our own countries.

In her lectures and seminars in North America and Europe, Helena Norberg-Hodge describes the ecological and social balance of traditional Ladakhi culture. In discussing how conventional development erodes that culture, she makes audiences more aware of some of the causes of problems in Western societies. She endeavors to inspire people with the notion that a more sustainable and humane way of life is achievable. Ladakh, she believes, can serve as a model and help show the way.

VANDANA SHIVA (1952–)
Global Militant

Operating out of her mother's cowshed in Dehra Dun in India's Himalayan foothills, Dr. Vandana Shiva and her research associates conduct science and technology assessments, focusing on conflicts over natural resources.

In 1981, when she established her Research Foundation for Science, Technology, and Natural Resource Policy (a heavy title, she admits, for a cowshed endeavor), she wanted to set up a research program not dependent on outsiders' money. She says the work of assessing new technologies forces one to recognize that there is almost always a conflict between technology's interests and those of ordinary people. Her heart is with the people, she says, and she vows to keep it there. "Ecologically and economically inappropriate science and technology can become causes of underdevelopment and poverty, not solutions to underdevelopment and impoverishment.[17]

From her cowshed headquarters, Shiva, a physicist, philosopher, ecologist, author, feminist, and global lobbyist, has initiated a worldwide rebellion against the genetic engineering of seeds, foods, and life forms by multinational corporations. The practice, she contends, deprives local farmers of their property rights and thwarts sustainable agriculture. A leading critic of today's reproductive and agricultural technologies, she calls her most recent book, which deals with the issue of intellectual property rights, *Biopiracy*.[18] "It is inadequately recognized that patents on life and intellectual property rights in the domain of life forms and biodiversity are actually an extremely sophisticated form of crude piracy" and are not in the public interest, she charges.[19]

The Southern countries' genetic richness provides much of the material being modified by genetic engineering and patented by Northern corporations. Even though Southern countries freely share these materials with the North, corporations can resell the modified materials to the countries of origin at high prices. India's neem tree is an example Shiva described in a 1996 speech she gave in Seattle to benefit the Edmonds Institute. For centuries the neem tree, because it disrupts reproduction in insects, has served as a natural pest-control agent; insects avoid anyone standing near it. Women traditionally have placed the tree's leaves among clothing and in seed storage containers. Even after centuries of such use, pests have not built up resistance to the neem tree's effectiveness. Using a run-of-the-mill chemical stabilization process, Shiva contends, an American transnational chemical company has obtained patents on pesticides and fungicides derived from the neem tree and has claimed monopoly rights. She notes that all told there are currently thirty-three patents on the neem tree held by Western corporations. The market price in India for the tree's seeds has increased by 1,000 percent over what it was twenty years ago, making the resource inaccessible to ordinary people.

In the same speech Shiva also cites the case of a wild herb, *Phyllanthus niruri*, that grows all over India. Seventy percent of that country's health care is based on herbal medicine, she says. For millennia the Indians have used this particular herb to cure jaundice and hepatitis, a fact confirmed in ancient Ayurvedic texts. Whereas Western medicine has been able to treat only the symptoms of hepatitis, this herb actually cures the disease by restoring liver functioning. In the 1950s and 1960s, she notes, chemical labs in India had worked out the chemical structure of the herb's active ingredients. Now, according to Shiva, the Fox Chase Cancer Research Center in Philadelphia claims that this herbal treatment for hepatitis is an invention needing patent protection.

In brief, such patents mean that researchers not working under a license belonging to the patent holder cannot use the genetic materials and techniques covered under the patent. Shiva, representing subsistence farmers—especially women—in Southern countries, is leading the effort to challenge this arrangement in Europe and the United States. For mil-

lennia, she argues, Southern countries have protected biodiversity on behalf of the world community and therefore hold prior rights. She has
compiled a four-page list of multinationals' patents for organisms taken
from tropical countries' soil for producing antibiotics and chemicals. For
instance, "Pfizer—without asking either the government or the people of
India whether the soil or microorganisms could be taken—holds a patent
for production of antibiotics from microorganisms taken from India's
soil."[20]

In *The Violence of the Green Revolution*, Shiva writes that between 1976
and 1980 wild plant varieties taken from Southern countries contributed
$340 million a year to the U.S. agricultural economy. And the cumulative
value of wild germ plasm to the U.S. economy—$66 billion—exceeds the
combined international debt of Mexico and the Philippines. However,
Southern countries have not shared in these profits.[21]

Shiva faults the American attitude claiming that biological diversity
is the common heritage of humankind. She calls this an excuse, which
allows industrial nations to steal without having it called stealing. In her
1996 Seattle speech, she observes:

> The relationship between the human species and other species has to be
> decided on ethical terms. It cannot be decided on commercial terms, but
> of course ethics is a barrier to free trade. And that is what industry has
> been so organized in removing. . . . It is time we recognize that we are no
> more ruled by governments, nor by our congresses, nor by our parlia-
> ments, nor by elected leaders—we are ruled globally by corporations.[22]

And in her 1993 book, *Monocultures of the Mind*, Shiva points out:

> The issue of privatization is increasingly becoming a threat to democracy
> and people's will, as the same scientists work for TNCs [transnational cor-
> porations], function on government regulatory bodies, and dominate sci-
> entific research. In this context it is up to citizens, free of TNC and
> government control, to keep public issues and priorities alive, and have a
> space for public control of the new biotechnologies.[23]

Shiva suggests that the North-South imbalance can be corrected only by
recognizing the local communities' contribution to biodiversity develop-

ment and replacing a bioimperialist regime with one based on biodemo-
cracy. Since 1985 she has written 20 books and published 150 articles in
technical and scientific journals presenting her views on these issues.

Indigenous cultures have always considered seeds precious resources.
Seeds have evolved over millennia through farmers—mainly women—
gathering the best, saving, storing, protecting, and testing them, and
maintaining the genetic base of food production. Even in the face of starvation,[24] the sanctity of seeds kept them off-limits as food. Since the green
revolution began in the 1960s, multinationals have been gaining control
over food production, acquiring seed companies, and patenting seeds. If
the seed is patented, farmers can no longer plant those saved from a previous year's crop without paying a royalty.[25] Between 1968 and 1988, for
example, Shell bought over sixty seed companies and Ciba-Geigy purchased twenty-six, and Imperial Chemical Industries (ICI) is now among
the top ten seed producers.[26]

Shiva characterizes Western patriarchal science and economic development as simply new aspects of colonialism. Western patriarchy's economic vision of "development," in her view, is based on the "exploitation
or exclusion of women, . . . on the exploitation and degradation of nature,
and on the exploitation and erosion of other cultures." This is why, she
says, women and indigenous people in the developing countries are
"struggling for liberation from 'development' just as they earlier strug-
gled for liberation from colonialism." She denounces Western-style development stemming from such patriarchal concepts as centralization,
compartmentalization, domination, and homogeneity as really being mal-
development, "which sees all work that does not produce profits and capital as non- or unproductive work."[27] Growth in this system, she notes,
relies on creating markets for things people do not really need. Labor
productivity for capital accumulation, she points out, differs sharply from
labor productivity for survival. Shiva's alternative model for development
would seek to restore balance in an integrated system of economics, politics, and technologies—a system that does not derive power for one of its
elements by crushing the others, but by empowering them all.

Shiva has few good things to say about the green revolution. Although in the short term it may have helped eradicate hunger, in the long
term it has not turned out to be the panacea promised by scientists for

the Southern Hemisphere countries. Instead, it has led to ecological deg-radation, cultural breakdown, and the destruction of forty centuries of sustainable practices. Shiva observes that "[t]he masculinist paradigm of food production which has come to us under the many labels of 'green revolution,' 'scientific agriculture,' etc., involves the disruption of the es-sential links between forestry, animal husbandry and agriculture, which have been the sustainable model."[28] She charges that the chemical-inten-sive, water-intensive, energy-intensive, and capital-intensive monocul-tural practices of the past two decades have caused erosion, pollution, deforestation, and serious damage to surrounding ecosystems, and have created deserts out of once fertile soils.

> Water-logging and salinity, micro-nutrient deficiency, toxicity and the depletion of organic matter are direct and inevitable consequences of a philosophy of agriculture guided by the modern patriarchal principle of profit-maximization. The recovery of soils can only take place through a philosophy which sees soil fertility, not cash, as agricultural capital . . . which puts nature and human needs, not markets, at the centre of sustain-able agriculture and land use. If soils and people are to live, we must stop converting soil fertility into cash and productive lands into deserts.[29]

Since 1970, Shiva asserts, the green revolution experiment has ru-ined land that could have produced much more food. A third of India has become wasteland. Half of Punjab, once known as India's wheat basket, now lies unproductive. Malnourishment haunts 60 percent of India's chil-dren.[30]

Shiva points out that both farmers and think-tank globalists now recognize the green revolution's failures. Many farmers no longer use the so-called miracle seeds. Now scientists offer genetic engineering as the answer. But, she notes, scientists can be as fallible as anyone, and they have usually prostituted themselves to economic interests. Furthermore, biological experiments and product testing are moving to the South—to escape bans, regulations, and public control in Northern countries—making the peoples of the South guinea pigs without their informed con-sent.[31] Biotechnology, she cautions in *The Violence of the Green Revolution*, may lead to disastrous results, simply hastening genetic erosion. She de-clares elsewhere:

We do not need genetic engineering to put nitrogen-fixing genes on maize and millet when women and peasants, for centuries, have used the more ecological option of intercropping maize with nitrogen-fixing beans, and millet with nitrogen-fixing pulses [for example, lentils and black beans].[32] It is not that nature is inadequate, only that corporations cannot make profits without manipulating nature. . . . Biotechnology corporations have merged with seed companies which are also producers of fertilizers and pesticides. The new seeds will be engineered within the old corporate control of Dow, Du Pont, Eli Lilly, Exxon, Merck, Monsanto, Pfizer, Upjohn, etc.[33]

Thus, Shiva laments, we are seeing the privatization of crop varieties that have been part of the common heritage of humanity for centuries—privatization by powerful interests seeking profitable monocultures that concentrate on a few high-yielding cash crops for export: hybrid wheat, corn, and rice. Monoculture's biological uniformity invites pests and diseases and depletes soil nutrients. Crop rotations, intercropping, and polyculture,[34] she points out, all traditionally practiced by women in Southern countries, simultaneously nurture soil fertility, create natural fertilizer, reduce vulnerability to drought, and control pests and diseases.[35] In other words, traditional, diversified cropping systems provide built-in protection. In these ancient practices, nature works quietly and invisibly. By contrast, monoculture, requiring irrigation and applications of chemical fertilizers and poisonous pesticides, presents a visible show. "Monocultures spread not because they produce more, but because they control more. The expansion of monocultures has more to do with politics and power than with enriching and enhancing systems of biological production."[36]

From the point of view of nature and poor peasants, Shiva avers, green revolution hybrid seed varieties (which cannot reproduce, but must be bought each year from seed companies, and which cannot function properly under natural conditions) were poor alternatives for increasing food production. They were useful only to corporations seeking new markets for seeds, fertilizers, pesticides, and irrigation systems. "Plants that have been displaced by plant improvement in the green revolution are pulses and oilseeds, which are crucial to the nutrient needs of people and the soil. Monocultures of wheat and rice . . . have also turned useful [and highly nutritious] plants into 'weeds,' as is the case with green leafy vegetables which grow as associates."[37] Industrialized agriculture and cash-crop export programs enhance men's incomes. For women, however—still expected to produce food crops for family survival—it means being pushed to less fertile plots of land father away from home.

✤ Shiva says a love of nature has always given her deep satisfaction. Believing that a profound understanding of the natural world could be gained through studying physics, she earned degrees in nuclear physics while also studying chemistry and botany. But she experienced "massive disappointment" in science as taught in academia. For example, "We were taught how to create chain reactions in nuclear material and we knew all about energy transformations, and so on, but nothing about the interaction of radiation with living systems."[38] In the mid-1970s, while working on her master's degree at the University of Guelph, in Ontario, she became aware of how uninformed nuclear scientists were, an awareness heightened when she worked in a nuclear reactor plant. She learned about radiation hazards not from her nuclear physics courses but from her physician sister. She warned her about its effects on human fertility and the unborn, and made her promise not to work there any more.

Troubled by basic questions regarding science, Shiva spurned a career in nuclear physics after she received her master's degree and enrolled in a doctoral program at the University of Western Ontario that offered a multidisciplinary curriculum. From that program, where she focused on science and technology policy issues, she acquired a doctorate in the foundations of physics.

Shiva had grown up in the Himalayan foothills of Dehra Dun, where her father worked as a forester. After being away at college for seven years, she returned home to find that "[t]he old forests were not there any more. The area had been converted to apple orchards. 'Growth centers,' as they call them in the vocabulary of development. . . . They put up a few shops and turn an area into an urban slum and call it a growth center."[39] She heard about the Chipko women and volunteered to help in an unstructured network called Friends of Chipko. Among other activities, she documented deforestation's environmental damage, and she and her colleagues prepared counter-reports to scientific papers opposing the

Chipko efforts. She believes the Chipko movement marked India's reawakening to ecological consciousness. Her involvement with the women- and nature-centered movement, she says, taught her to think in a feminist and ecological way, and led to ten years of research focusing on conflicts over natural resources.

Before organizing her own independent operating foundation in Dehra Dun, Shiva had engaged in research for government agencies, first at the Indian Institute of Science and then at the Indian Institute of Management, where she focused on science policy. Since establishing her independent research facility, she has surfaced as a worldwide militant presence and activist role model through her writings, lectures, lobbying, and participation in international forums. Named among *Utne Reader*'s visionaries who will change our lives, and the 1993 recipient of the Right Livelihood Award, she appeared as a featured speaker at the May 1996 Global Teach-In 2 cosponsored by a pair of nonprofit NGOs—the International Forum on Globalization (of which she is a founding member) and Public Citizen. One of the forty thousand women at the 1995 UN Women's Conference in Beijing, Shiva, as one of the chairwomen of WEDO,[40] had earlier helped Chairperson Bella Abzug organize the 1991 World Women's Congress for a Healthy Planet. Shiva felt this Miami conference of fifteen hundred women from eighty-three countries was the most inspiring event she had participated in for a decade. While at the 1992 Earth Summit (UNCED), Shiva recalled the Miami women's conference that had taken place a few months earlier: "It was very different from UNCED, where 99 percent of the delegations just sit there without any commitment. . . . They are 99 percent men and 99 percent of them have orders not to do anything.

"At the official level, UNCED delegates seem bent on killing a global environmental commitment. That is the reading I am getting, and the U.S. position is the best example of that."[41] The UNCED Convention on Biodiversity, crafted under U.S. domination, falls short in that it leaves the door open for Northern countries to patent genetic materials. It does, however, give countries the right to charge for access to their genetic resources, and it permits them to enact legislation outlining terms for bioprospecting contracts.[42]

Meanwhile, in addition to her research Shiva designs action campaigns on specific issues and files lawsuits against corporations. Another project she has initiated is the organizing of community-based seed banks in India. Already, the program has preserved hundreds of varieties of beans, rice, and millet.

"Why is it," Shiva asks, "that women sense destruction faster and are more persevering in the struggles against destruction? Why do they carry on when everyone else is cynical and hopeless? The reason is that women have a distinctive perception of what life is, a sense of what is really vital, which colours their view of what is at stake in the world."[43] She posits in her writings and lectures that hope for the survival of nature and the peoples of the South rests on applying women's ideas and actions passed on from earlier generations.

FLORENCE WAMBUGU (1953–)
Cultivates Food Security for African Farmers

Unlike half of Africa's professionally trained scientists, who choose to work outside the continent, Dr. Florence Wambugu of Kenya, concerned with eliminating hunger in Africa, applies her knowledge of botany, agronomy, plant pathology, and biotechnology to improving food crops at home.

In Kenya and other African countries, converting diversified acreage to monocultural cash crops for export undermines food security and, Wambugu observes, "often is the result of lack of power and influence in decision making on the part of women, who are the backbone of agricultural production. This results in hunger and malnutrition in most African countries." The concentration on cash crops for export has left African family farmers (mostly women) with only small parcels of land on which to raise subsistence crops to support their families. She notes that "most men view development and security of a country in terms of export strength, an army, cash, cars, and equipment. Women's view of security, growth, and development is mainly based on quality of life as it affects especially the ability to provide for the family in terms of children's needs—food, clothing, health, and education." Wambugu believes that food, not money, constitutes a nation's real security. She focuses on improving the lot of Kenya's, and Africa's, food-producing farmers, choosing the sweet potato as her specialty. She says of her choice:

> I was particularly interested in sweet potatoes because they are grown by small-scale farmers and support the lives of women and children. The sweet potato is a drought-resistant crop that grows in poor soils without fertilizers and pesticides. It is rich in carbohydrates and nutrients, but it is also susceptible to a variety of diseases, among them the feathery mottle virus spread by aphids. This disease causes the plant to degenerate, producing smaller leaves and potatoes and normally reduces yields by half. In spite of this crop's food security and great potential, very little research had been done.

Wambugu undertook the challenge of developing a disease-resistant variety. Ultimately, her work became a model for all research on sweet potatoes in Kenya and Uganda.

In 1978, after receiving her bachelor of science in botany from the University of Nairobi, she spent four years working with the Kenya Agricultural Research Institute (KARI). Her first assignment in their plant quarantine station was checking tissue culture samples from abroad to detect invader diseases that might jeopardize the country's food supply. To sharpen her expertise, KARI sent her to a month-long training program on tissue culture technology given by the International Institute of Tropical Agriculture in Nigeria. Later, this institute gave Wambugu an award for successfully establishing the tissue culture lab at KARI "in support of root and tuber crops germplasm improvement, introduction and exchange."

During the 1980s Wambugu seized every opportunity to broaden her understanding of the cell and tissue culture of plants by participating in short courses, seminars, and workshops at various universities in the United States, study tours on various crops in the United Kingdom and Canada, and international conferences. At several national and international meetings, she herself served as a trainer.

Wambugu considers her experience at KARI to be a turning point in her career. While there, she learned about international agencies like the United States Agency for International Development (USAID) and the International Potato Center (CIP). Of the latter she recalls:

CIP for the first time was importing potatoes from Peru, where they originated thousands of years ago. Potatoes from the Netherlands and other European countries had been introduced to Kenya during colonial times, but the new strains arriving from CIP's gene bank in Huancayo [Peru] promised new vigor and genetic diversity that could ensure better food supply in Kenya.

She took time out from her KARI job in 1982 to obtain a master's degree in plant pathology from North Dakota State University, where she specialized in the control of viruses in potatoes. Returning to KARI, she undertook research on root and tuber crops in the institute's plant pathology section.

Up to that point her efforts at KARI to reduce losses in sweet potato crops had failed. But her determination to solve the problem drove her to seek further scientific knowledge. She proposed a time-splitting arrangement between KARI and the University of Bath in England, allowing her between 1984 and 1991 to pursue her doctorate on the epidemiology of the sweet potato. By this time, the institute had promoted her to the position of senior research officer (pathologist) and coordinator of plant biotechnology research. On completion of her doctoral work, both KARI and Horticultural Research International cited her in 1991 as an exemplary Ph.D. candidate for her outstanding dissertation on virus diseases of sweet potatoes. But she still had not found the answer she sought.

✽✻ Wambugu's fascination with plants stemmed from her girlhood on a small farm in central Kenya. Her mother, she says, "like many African women, was solely responsible for raising her family of six children while Father was away employed on a farm owned by Europeans. She coaxed a small piece of land into yielding enough money to feed, clothe, and educate all of us." Reflecting on her childhood experiences growing vegetables with her mother, Wambugu says that their farm had been her best research laboratory. Even though they did not use the technical terms, they actually practiced agronomy, plant pathology, and breeding in growing their food crops without agricultural chemicals. The misuse of pesticides, even though they are applied mainly to cash crops for export, has always been one of her biggest concerns. "Their overall use in Africa has not made a major impact on food production. These chemicals in many instances have posed an environmental hazard as well as a health risk to many African farmers."

Wambugu's mother encouraged the children to pursue education, sending her daughter to a girls' secondary boarding school, "which was a major step in those early days [1970–74], considering that girls were often left at home to help with household chores while boys stood a better chance of furthering their education." In high school, chemistry and biology caught her interest, stimulating ideas for a career in agricultural research.

✹ Wambugu's defining break came in 1991, when USAID selected her to receive the first of its new postdoctoral fellowships in the uses of biotechnology to improve root and tuber crops for African farmers. The agency and the Monsanto Company jointly funded the three-year fellowship, which allowed Wambugu to study genetic engineering at Monsanto's Life Sciences Research Center in Missouri. Wambugu had never been exposed to genetic engineering in practice, but she quickly mastered the technology.

Her research in Missouri focused on using biotechnology to create a disease-resistant sweet potato by transferring genes at the cellular level. Wambugu recalls that she

> achieved a breakthrough by splicing the gene that triggers virus resistance and placing it into the genome [a full set of chromosomes]. This pioneering work sets an example for future collaboration between developed and developing countries. We developed a virus-resistant sweet potato which will become a commercial product. . . . The main gene that was used for the development originated from the public domain, as it was sequenced and cloned using funds from the International Potato Center, and as such it is not patented. Monsanto Company has waived royalty claims for the other supportive genes that were used to develop this technology under our current agreement. Kenya and all of Africa will have a free right to operate this technology.

The technology is being transferred to KARI in Kenya, where extensive field trials on four varieties of transgenic disease-resistant sweet potatoes are now under way.

❧ During her years at KARI, Wambugu, a divorced mother of three, helped found a primary school for the children of the institute's staff. She served as treasurer during its first year in operation in the mid-1980s. In the early 1990s, while conducting her postdoctoral research at Monsanto, and as part of a cultural exchange program, she visited and gave talks to elementary and high school students in Missouri about her efforts to alleviate world hunger. She also presented her work at various American universities, government agencies, and scientists' conferences.

Since 1994 Wambugu has been director of *Afri*Center, one of a network of six worldwide centers supported by a nonprofit group, the International Service for the Acquisition of Agri-biotech Applications (ISAAA). The service's mission "is the transfer and application of appropriate technologies for sustainable agricultural production in developing countries, including [those of] Africa." The organization facilitates the transfer of technology from developed to developing countries by matching worthy projects with funders. Current ISAAA *Afri*Center projects to improve the living standards of rural families, especially Kenya's resource-poor subsistence farmers, include development of a multipurpose tree variety, better banana production, a virus-resistant line of maize, overseeing the sweet potato field tests, and setting up the process for transferring the virus-resistant varieties to African institutions.

Wambugu, as director of *Afri*Center, is also developing the infrastructure needed to handle technology transfers, such as a biotransformation laboratory, a biosafety regulation system, and a training program and guidelines for handling transgenic field trials. She also deals with issues relating to intellectual property rights and patents.

She has written or collaborated on more than thirty-six scientific papers for technical periodicals in a number of countries. Since the International Potato Center has no other biotechnology expert in Nairobi, she also conducts training courses for Kenyan and Ugandan scientists on plant pathology and biotechnology.

What role does Wambugu believe women should play in solving the earth's ecological predicament?

Women need to inspire and support each other to get professional training, allowing them to move into strategic positions where they can be key players and influence major policy decisions that affect the world, nations, and regions. This way issues such as food security and rural poverty can receive priority attention over equipping the army. We also need to encourage women in the younger generation to become active in politics rather than being passive observers. Women are more sensitive to ethical, ecological, and cultural issues, and they try to take responsibilities. They are more for conservation of natural resources, environmental protection, and the attendant long-term benefits.

Wambugu is one of the few African scientists working to enhance food crops. Though cash crops for the export business have led to underdevelopment in the continent's food production, she anticipates that her work will eventually help ensure that African children can grow up without experiencing hunger or malnutrition.

PETRA KELLY (1947–1992)
Green Party Pioneer

In the fall of 1992, the brutal death of Petra Kelly, cofounder of the German Green Party, stunned her colleagues and shocked the world. The police immediately alleged a murder-suicide at the hands of her Green Party colleague and life companion, Gert Bastian. But could they have been assassinated? We may never know, since no official (or even informal) investigation took place, even though puzzling circumstances pointed to other possibilities. For one thing, Kelly had received threats on her life, leading the police to refer to her as "an endangered person." And there were other bewildering pieces of evidence.

But who was Petra Kelly, and what did she do that might have provoked hatred?

✎ Born in Günzberg, Bavaria, Kelly learned nothing about the Holocaust at her convent school. She heard about it only after her family came to the United States, where she spent her adolescent years. At American University in Washington, D.C., where she majored in world politics and international relations, two professors inspired her to become an anti-authoritarian critical thinker. During her years in the United States, from

1960 to 1970, she learned about nonviolent protest by participating in the civil rights and antiwar movements. She learned about politics in 1968 as a volunteer in Robert Kennedy's and then Hubert Humphrey's political campaigns.

After graduating cum laude from American University, she returned to Europe and obtained her master's degree in political science and European integration at the University of Amsterdam. She then took a job as an administrator with the European Economic Community; for almost a decade, she did trailblazing work on social, environmental, health, and education issues. As a volunteer she organized local citizens' initiatives against nuclear power plants and airport expansions. Police arrested and fined her several times for leading sit-ins at military installations.

As early as 1968 she and her grandmother had participated in the nonviolent resistance movement when Warsaw Pact forces occupied Prague. (The two of them were placed under house arrest in a hotel.) Two decades later the people of the former Czechoslovakia demonstrated the power of nonviolent protest in the successful "velvet revolution." During the 1970s Kelly also joined nonviolent protest marches in Germany and Ireland against nuclear plants and weapons and against police terrorism and the criminalization of pacifists. She described the spiritual power of her approach this way:

> The power of nonviolence arises from what is deepest and most humane within ourselves and speaks directly to what is deepest and most humane in others. Nonviolence works not through defeating the opponent but by awakening the opponent and oneself through openness. . . . Nonviolence is a spiritual weapon that can succeed where guns and armies never could.[44]

Though raised a Roman Catholic, Kelly left the church, saying she found deeper meaning in Eastern spiritual traditions, having learned about Buddhism from His Holiness the Dalai Lama—with whom she had forged a close alliance—and from a Tibetan foster child she supported.

During her twenties, as a loyal supporter of West German Chancellor Willy Brandt (who opposed nuclear arms and nuclear energy) and of the West German Social Democratic Party, Kelly plunged into politics.

Becoming disillusioned in 1974 when nuclear advocate Helmut Schmidt became its leader, she soon resigned from the party.

She then persuaded a few friends to join her in creating a new political party modeled on ecological, antimilitary, and feminist principles—an "anti-party party, or counter power," as she called it. In 1979, the West German Green Party came into being.

About that time, the North Atlantic Treaty Organization announced plans to deploy the "cruise" and Pershing II missiles in Europe. Alarmed by the increased threat of nuclear war posed by this plan, the Green Party organized citizen protests against the missiles. European sentiment against deployment escalated. On 10 October 1981, Petra Kelly, as Green Party leader, delivered an electrifying speech in Bonn to a rally of 250,000 peace marchers. The rally and her speech made global headlines. Overnight, she found herself in the limelight.

In 1982, the German Greens' soaring popularity led to their winning 27 seats (out of 498) in the German Parliament—the first parliament in the world to have a Green contingent. Kelly was elected one of the Greens' three parliamentary speakers. During her eight years in Parliament, Kelly used her office to advance ecological, human rights, and feminist causes, to eliminate arms exports, and to lead protests against nuclear weapons. She believed the strategy of nuclear deterrence to be a form of collective hysteria and blackmail. In November 1986, she and some friends sat in the rain in nonviolent protest at the NATO building in Hasselbach, the town where the cruise missiles were to be deployed. Police hauled them away, calling their actions "reprehensible." Kelly notes, "According to judgments handed down by West German courts, Tolstoy, Gandhi, Jesus and Martin Luther King all acted in a reprehensible manner" (*TG*, 50).

Described by her associates as charismatic, intense, driven, articulate, fast-talking, and a prima donna, she followed a grueling schedule. Though often urged by her friends to slow down, the Dalai Lama once prompted her instead to keep going, saying that he would do her meditating for her.

Kelly, as an advocate of human rights, defended refugee immigrants in Germany from attacks by neo-Nazis. Recognizing the wisdom of traditional people, she promoted the interests of indigenous and oppressed groups around the world: First Americans (Indians), African and South American native people, Australian aborigines, the oppressed people of Tibet, and the Tiananmen Square demonstrators of 1989 in Beijing. She wrote many articles on ecology, feminism, human rights, disarmament, and Hiroshima. She also wrote several books, including *Fighting for Hope*,[45] her personal manifesto. And she coedited *The Anguish of Tibet*,[46] a collection of articles about the ecological, cultural, and physical destruction of that country by the Chinese Communist government. Since it invaded Tibet in 1950, over a million Tibetans have been murdered, six thousand monasteries destroyed, forested hillsides stripped, and most of the wildlife killed.

If any single group can claim credit for the collapse of the Berlin Wall, the German Green Party and Petra Kelly deserve to be at the top of the list. The German Greens was the only party to have continuous close contacts with dissident grassroots citizen action groups in East Germany and Eastern Europe during the 1980s. According to Kelly, strong-minded women were among the leaders of the 1989 East German revolution. Since parliamentary immunity allowed her to cross the border without being searched, Kelly had often transported office supplies to the fledgling democracy movement in East Germany.

Fifty years of industrial pollution had reduced the Eastern European environment to a wasteland. Kelly observes:

> The environmental outrage of the people was so high that it helped topple governments in several countries. . . .
> The tearing down of the Berlin Wall, the collapse of the communist regimes in Eastern Europe, and the thawing of the Cold War brought the promise of a large-scale redirection of society's resources and priorities. We heard so much talk of peace dividends, from converting the permanent war economies, to ones geared toward addressing social and environmental problems. (*TG*, 111, 115)

During the 1980s the German Greens had discussed with Eastern European dissidents their shared dreams for an ecological, nonviolent, and gradual transformation after Communism toward a model society. They envisioned a demilitarized, denuclearized, nonaligned Eastern Europe.

For that reason, the German Greens opposed a too rapid German reunification.

On 6 December 1989, as part of the planning for the transition, Kelly took the Dalai Lama into East Berlin to participate in a round table of activists forging a new German constitution embodying pacifism, ecology, social justice, and feminism. She described the meeting as momentous. The participants vowed that the new East Germany would recognize Tibet in defiance of China.

In 1990, Kelly also escorted the Dalai Lama to Prague to visit President Václav Havel and to celebrate Czechoslovakia's first free elections in more than four decades. Havel made the Dalai Lama his first official guest.

Unfortunately, the German Greens' dreams of a model ecological society in East Germany crashed when Western bankers, businessmen, and politicians surged into Germany following the disappearance of the Iron Curtain, imposing their blueprint for what they considered an ideal capitalist social order. Kelly remarks:

> The shift to market-oriented economies took place with virtually no regard for the environment. Western businesses and governments made it clear that they are no more interested in ecological transformation in Eastern Europe than at home. . . . It is tragic that the extraordinary opportunities that came with the collapse of communism have been squandered. . . . I still hope that an ecological model of economic development—an alternative to repressive state socialism and to aggressive capitalism—can emerge somewhere in Eastern Europe, perhaps in Poland, Hungary or Czechoslovakia. (*TG*, 112, 113, 119)

In her home country, Kelly feared the resurgence of Nazism and a militarist German nationalism. In October 1990, during the first joint session of the two German parliaments, the German Greens pushed unsuccessfully to mention the Holocaust in the preamble to the Unification Treaty. She points out:

> Many of us fear that Germany's economic power will lead to military power, and a future militarist German nationalism with a nuclear capacity is something we must speak about realistically. Germany and Germans

caused and were entwined in two terrible wars and attempted to eliminate the Jewish people from the face of the Earth. We must practice self-restraint, such as the ban on all arms exports, eliminating the military-industrial complex, no war financing, instituting a Peace Corps, and large-scale investment for ecological reconstruction in Eastern Europe and the South. (*TG*, 97)

Since reunification, according to Kelly, neo-Nazis have killed more foreigners than the total number of people killed trying to cross the border during the years of Communism. And more and more acts of anti-Semitism have erupted all over Germany.

To create a generation of peacemakers, she recommended a peace studies education program to develop practical methods for conflict resolution and to foster nonviolent social defense systems. "Human dignity is a fundamental value in peace education," she wrote. "This means moving away from the emphasis on competition, achievement, strength, power, profit, and productivity. . . . Peace studies can, I hope, become truly peacemaking—helping develop an ethic of reverence for life on this Earth, a planet that has no emergency exit" (*TG*, 56).

In 1990 the German Green Party failed to win enough votes to retain its seats in the lower house of parliament, the Bundestag. Kelly attributed the failure to party power struggles and feuds, jealousies, distrust, and dissension. "We have taken on all the bad features parties usually display—including financial scandals, a credibility crisis, and an inability to resolve problems in a constructive way" (*TG*, 123). She had always held the top position in the party, and she resisted the principle of leadership rotation. This did not sit well with her party colleagues, some of whom also resented the worldwide media attention she received. Nonetheless, the passionate concern with which she articulated the Green perspective to the international media seduced even hawkish, conservative journalists. For example, one of them was so impressed with her performance on *Meet the Press* in 1983 that he told her afterward he wished she were on his side of the issue.

Fearing that the German Green Party might not survive,[47] she hoped others would learn from its mistakes. She urged Americans to initiate Green politics at every electoral level.[48] She believed that Green politics must begin at the grass roots; politics from the top is usually corrupt, and the top will change only with sufficient pressure from below. She also perceived an even greater exigency: "Politics needs spirituality," Kelly wrote.

> The profound political changes we need in order to heal our planet will not come about through fragmented problem solving or intellectual analyses that overlook the deepest yearnings and intuitions of the heart. . . . As we begin to cultivate a rich inner life and experience our connection with all of life, we realize how little of what society tells us we need is actually important for our well-being. . . . Green politics must address the spiritual vacuum of industrial society. (*TG*, 40, 41)

Thanks largely to the group's efforts, Germany became the most environmentally conscious of the world's industrial countries.[49]

Kelly spent the last few weeks of her life on the publishing details surrounding what would be her last book, *Thinking Green*—a collection of her writings, speeches, and interviews. Her publisher, as well as the environmental writer Peter Matthiessen, who wrote the foreword, described having upbeat, enthusiastic communications with her about the project shortly before the shooting. She reportedly planned to run for political office again in 1994.

The puzzling circumstances surrounding her murder included the discovery of an inconsequential letter to an attorney still in Gert Bastian's typewriter in their apartment—a letter that was interrupted in the middle of a word. Furthermore, the electric typewriter was still running when the bodies were discovered. Police pointed to gunpowder on Bastian's hand as positive evidence that no third party was involved. However, skeptics have pointed out that any well-trained secret service or other operative would have known how to fake all the circumstances surrounding the tragedy.

Called by her publisher one of the twentieth century's great peacemakers, Kelly received the 1982 Right Livelihood Award. And *Time*, in reporting her death, referred to her as "the world's best-known environmentalist."[50] Her interests encompassed ecology, peace, disarmament, nonviolent civil disobedience, a nonnuclear society, and the human rights of women, immigrants, and minorities—all on a global scale.

Petra Kelly urged women to seek political office to change policies from those of death to those of life. She said she wanted shared power *with* others to replace patriarchal power *over* others. "Masculine technology and patriarchal values," she wrote, "have prevailed in Auschwitz, Dresden, Hiroshima, Nagasaki, Vietnam, Iran, Iraq, Afghanistan, and many other parts of the world. . . . Feminists working in the peace and ecology movements are sometimes viewed as kind, nurturing Earth Mothers. . . . [B]ut we are not weak. We are angry—on our own behalf, for our sisters and children who suffer, and for the entire planet—and we are determined to protect life on Earth" (*TG*, 11, 12).

HAZEL WOLF (1898–2000)
Models Voluntary Simplicity

Newly arrived in Seattle from Canada in the early 1920s, Hazel Wolf approached the YWCA swimming pool occupied only by a few African-American women. Having been advised by the attendant that it was "Negroes Day" at the pool, Wolf, a superb swimmer, obtained consent to swim anyway. Before diving in, she asked (and received) the black women's permission to join them in the pool.

Defying rules she deems unjust governed Wolf's lifestyle. As a child early in the twentieth century, she protested doing dishes unless her brother also took a turn.

A reformer, feminist, organizer, and activist with a multitude of causes, Wolf was a late-blooming environmentalist. Although she was an outstanding athlete as a young woman (once running the hundred-yard dash in eleven seconds) and had engaged in outdoor adventuring all her life—hiking, canoeing, camping, kayaking, snowshoeing, mountain climbing, sailing, and swimming—her activism before she entered her sixties focused on the world peace movement and on social, racial, and political reform. While on welfare herself, she became a union organizer within the Works Progress Administration (WPA) during the Great Depression and even joined the American Communist Party for a period because no other group then championed unemployment insurance.

Born of an American mother and a Scottish seaman father (who died when she was small), Wolf had grown up in poverty before immigrating to Seattle from Victoria, B.C., in 1923 with her small daughter. Eventually she settled into a position as legal secretary for a civil rights attorney, assisting him for twenty years with cases involving social reform issues.

To pacify a hiking companion, but with little enthusiasm, Wolf joined the National Audubon Society at age sixty-two shortly before she retired from her secretarial job. Urged by the same friend to accompany her on a birdwatching field trip, she found herself enchanted by the behavior of a brown creeper. Watching the bird work its way up the trunks of one tree after another—always picking away for insects, up, up, up, never down—she said, "I felt protective toward this wonderful bird that knew where its food was, worked hard to get it, and maintained its lifestyle."

She soon became secretary of the Audubon Society's Seattle chapter (where women far outnumber men), a position she held for over thirty-two years. During those years, she instigated the formation of more Audubon chapters than anyone else in the national organization's history—twenty-three of them in Washington State alone, and another in her birthplace, Victoria, B.C. Anyone standing in line or occupying a seat next to Wolf received her pitch, often successful, to join Audubon.

Beyond recruiting and organizing members within just one group, Wolf's concern for the local and global environment motivated her to become an effective coalition builder among groups—with American Indians, with Nicaraguans, with poor communities, and with communities of color. She demonstrated that coalitions strengthen the collective influence of groups working on environmental and social issues. Wolf said that organizing new Audubon chapters and forming various coalitions were the most personally rewarding of all her activities. "The most valuable and interesting thing I've done was to organize a Conference of Native Americans [from Washington, Oregon, and British Columbia] together with environmental organizations, cementing associations which endure to this day." In making arrangements for the conference, held in Seattle in 1979, Wolf visited twenty-six tribal leaders in Washington State. The meeting brought together conservationists, farmers, and American Indians to protect the Columbia River from further dam building. Wolf had researched the issue, studying in detail the environmental

PHOTO: FRED FELLEMAN

HAZEL WOLF KAYAKING AT NISQUALLY REACH NATIONAL WILDLIFE REFUGE.

impact statement for the proposed expansion of the irrigation project. She wrote a comprehensive article about it for *Outdoors West*—an environmental newsletter, published by the Federation of Western Outdoor Clubs (FWOC), of which she had been editor for seventeen years—presenting the pros and cons of the water project. Eventually, the government dropped the expansion plans.

Wolf's brief affiliation with the U.S. Communist Party during the Great Depression haunted her amid the hysteria of the McCarthy era in the 1950s, when men and women—in violation of their civil rights—lost their jobs for refusing to take a loyalty oath. In a 1990 speech to high school students about the McCarthy period, Wolf observed:

> Loyalty or disloyalty is a very private matter because there is no way, with or without oaths, that the secrecy of one's inmost thoughts can be known. Therefore, it is obvious that the sole purpose of the loyalty oath was to intimidate people into suppressing their ideas about the Cold War campaign. Many refused to take the oath as a matter of principle, claiming that membership in the Communist Party was their right as citizens of a free country, just as was membership in the Democratic, Republican, or any other political party.[51]

An unemployed single mother during the 1930s, she had joined the party because of its crusade to gain unemployment assistance for jobless workers, to fight hunger, homelessness, and racism, and to protect women's rights. In 1955, thirteen years after she had quit the Communist Party out of boredom, the Immigration and Naturalization Service (INS) arrested Wolf, an "alien," charging her with "attempting to overthrow the government with force and violence." One of five persons in Seattle similarly accused, she spent half a day in jail working on a jigsaw puzzle before friends bailed her out. But in the decade following she faced repeated deportation attempts by the INS. One time, with agents close behind, she took off in her car on back roads and hid out. Her employer, the civil rights attorney, succeeded eventually in having her case (which had reached the Supreme Court) dismissed. It took fifteen years. She became a citizen in 1970.

As a young woman Wolf had not completed her formal schooling in Canada; later she earned a high school diploma in Seattle. She also enrolled in college on a scholarship but soon decided she could better educate herself through reading on her own. She said she took college courses only as needed. For example, when she was ninety-one she took a Spanish course at a Seattle community college in preparation for her fourth trip to Nicaragua (she had obtained an international press pass to observe the country's 1990 elections).

Wolf first visited Nicaragua in 1985 as part of a U.S. group invited by the Nicaraguan Association of Biologists and Ecologists. After getting to know many Nicaraguans, hearing about their problems and listening to their views, she returned home to oppose the official U.S. policy of supporting the right-wing Contra rebels. Following a speech on the subject at a Republican luncheon, several persons told her privately that they appreciated hearing the other side and getting information they had not known before. Wolf said that we could learn from the Nicaraguans' approach to the environment. For example, the Sandinistas nationalized all the rain forests when they took power in 1979. The new government also banned the use of DDT, and the country's scientists developed a breed of wasp that feeds on boll weevil larvae, two developments that reduced pesticide use by 50 percent. These and similar actions meant loss of business for U.S. timber and agricultural chemicals companies. The Nicaraguan Association of Biologists and Ecologists presented Wolf with a citation etched into a mahogany plaque with a burning-hot stick. She celebrated her ninetieth birthday by conducting a fund-raiser for medical supplies for a Nicaraguan hospital.

In 1989 Wolf represented Audubon at still another conference in Nicaragua, The Fate and Hope of the Earth. The meeting drew seventeen hundred people (many representing indigenous groups) from seventy nations around the world. Wolf and President Daniel Ortega both spoke at the final plenary session. Considered their special saint by the Nicaraguans, Wolf received a standing ovation when she ended her talk by saying, in Spanish, "The people united will never be defeated." In 1994 she again visited Nicaragua to help set up a model health advisory group.

In addition to her organizing and recruiting activities, Wolf crusaded for environmental issues—testifying at hearings, picketing at nuclear waste dump sites, writing articles, lobbying office holders, participating in environmental rallies and protests, and urging others to lobby and to register to vote. As the editor of *Outdoors West*, she prodded readers to take action on issues of current concern. In 1996 she planned the program for the sixty-fifth annual convention of the Federation of Western Outdoor Clubs. A regular attender of environmental conferences around the country, she often succeeded in spontaneously gaining the microphone during a plenary session to plead support for a particular cause. During the 1996 convention of the National Audubon Society in Washington, D.C., she participated in a Capitol Hill rally and lobbied senators and representatives to support environmental laws under attack in the 104th Congress.

Much in demand as an environmental speaker, Wolf captivated audiences with her wit and storytelling skills. When thirty learning-disabled teens in an Alabama classroom wrote to her after their teacher read aloud a *USA [Today] Weekend* cover feature about her,[52] she traveled to Alabama and gave a talk not only to that class but to others. Young children often asked her if she had a boyfriend. She told them no, but she was looking for one who liked to cook (she did not) and to let her know if they found someone. When asked if she traveled alone, she quipped, "No. Never. The planes, trains, and buses are always full of people."

In addition to *USA [Today] Weekend,* Wolf had been featured in many regional and national newspapers and magazines, including the *Seattle Times* and the *Seattle Post-Intelligencer, Audubon* magazine, and the politically conservative *Seattle Daily Journal of Commerce.* Studs Terkel included her in his oral history collection *Coming of Age.* In 1978, for helping put together a series of books about mammals, butterflies, and wildflowers of the Northwest, she received one of the Sol Feinstone Awards bestowed by the State University of New York. In 1985 the National Audubon Society recognized her efforts with its Outstanding Conservationist Award. And in 1997 she received one of Chevron's Conservation Awards, as well as the Annual Audubon Medal, an honor bestowed previously on such notables as Rachel Carson, Jimmy Carter, Edward O. Wilson, Ted Turner, and Robert Redford.

Wolf lived alone on Social Security in a small rent-subsidized apartment in Seattle. Over a dozen plaques and recognition citations decorated one wall. Hornets built a huge nest just outside her screenless living-room window in 1996. When her building manager suggested she let him spray it with an insecticide, she shouted, "Over my dead body. Those little guys work too hard!" She then trotted off to the library to research hornets and find out how they produce their paper dwellings.

Wolf had no medical coverage and no television set, and she gave up her car when she was ninety-four. But she used a word processor and printer for her editorial work and correspondence. She had one daughter, five grandchildren, five great-grandchildren, and two great-great-grandchildren. None of her grandchildren or great-grandchildren was an environmentalist, she said, "but then, none of them is yet sixty-two as I was when I became hooked."

In recent years Wolf had concentrated on extending the reach of environmental organizations beyond their largely white, middle-class membership. The biggest obstacle was "trying to get Seattle Audubon to broaden its focus beyond the protection of birds, other wildlife, and their habitats." She cofounded the Community Coalition for Environmental Justice (CCEJ) with members of poor and minority communities in the Seattle area. It is a model project to clean up industrial pollution in low-income neighborhoods, where industry most often sites toxic landfills and incinerators. Pollution, an environmental issue for organizations like Audubon, is a health issue for poor people and members of minority groups. Wolf believed these two perspectives on a shared problem form the bridge for the groups to coalesce. The Environmental Protection Agency awarded CCEJ a $75,000 grant to teach people how they can improve their indoor air quality.

Wolf felt that regularly spending time in the natural world kept her healthy and relaxed and prevented her from burning out. Still camping, swimming, and kayaking as she approached her hundredth birthday, she said she wanted to be buried in a kayak. She hoped a large crowd would attend her funeral so that it could be a major fund-raiser for a worthy cause. In 1996 Wolf declared she planned to live until the year 2000, so that her life span would have touched three centuries. "Then I'm going," she says. And she did.

On her ninety-eighth birthday the governor of Washington entertained a throng of environmental leaders and other dignitaries at a party in Wolf's honor and issued a proclamation declaring her birthday, the tenth of March, Hazel Wolf Day. Furthermore, during its June 1997 commencement exercises, Seattle University conferred on Wolf—then ninety-nine—an honorary doctor of humanities degree. Her Web site address (created by a friend) was

http://www.wolfnet.com/~brainkidHazelWolf.html

What gave her hope that humans would be able to save the planet's ecosystems? Wolf said for one thing the gregarious nature of humans—the desire to band together in order to survive. And although Wolf was an atheist, she pointed out:

> Almost every sacred book has as its golden core the concept of love for each other. . . . The citizens of the world will resolve our current ecological predicament the same way they have always resolved problems: they will organize. The chief enemies are greed and ignorance. Education will destroy ignorance, and even the greedy *may* finally realize they cannot survive in a polluted world. But without the leadership of women there will be no transformation.[53] The empowerment of women is of utmost importance.

CHAPTER TEN

A river seems a magic thing. A magic, moving, living part
of the very earth itself—for it is from the soil, both from
its depth and from its surface, that a river has its
beginning.

LAURA GILPIN
The Rio Grande, River of Destiny

Thinking Like a River

MARJORY STONEMAN DOUGLAS (1890–1998)
Champions the Everglades

In 1941, when Marjory Stoneman Douglas—already an accomplished writer—was six months into drafting a novel, an editor from the Rinehart publishing firm approached her about doing a book on the Miami River. She told him the river was only an "inch long" and one could scarcely do a whole book on it. "But when a publisher visits your house and asks you to write something, you don't let him go casually. I suggested that the Miami River might turn out to be part of the Everglades. . . . I knew it was connected to the Everglades. I can't pretend I knew much more than that." The editor suggested further investigation. "There, on a writer's whim and an editor's decision, I was hooked with the idea that would consume me for the rest of my life. . . . A great release of energy seemed to launch me into my most ambitious and important project. The Everglades book, no doubt my best writing, was a product of this personal renewal."[1]

The Everglades ecosystem, not a river in the usual sense, was originally a 110-mile-wide shallow wetlands, a complex natural water purification system. It carries a creeping sheet of rainwater south on an imperceptible slope from Florida's Lake Okeechobee through the Everglades to Tampa Bay and the Gulf of Mexico. Prior to human tampering, it required almost a year for water to complete the journey. Lake Okeechobee is fed by the Kissimmee River, which carries rainwater from the watershed farther north into the lake. Evaporation during the flow through the Everglades produces rain clouds, some of which drift north to repeat the cycle. South Florida's drinking-water supply depends on the functioning of this ecosystem.

The Everglades, home to fifty-five endangered and threatened native species and the wintering or breeding ground for millions of migrators, lie at the juncture of the tropical and temperate zones. Saw grass, with its long, pointed and serrated leaves sharp as saw blades, is the best-known natural feature of the Everglades. Really a sedge rather than a grass, and one of the oldest green growing plants in the world, it inspired Douglas to entitle her book *The Everglades: River of Grass.* "It stretches as it always has stretched, in one thick enormous curving river of grass, to the very end. This is the Everglades. . . . [T]his is the greatest concentration of saw grass in the world."[2] Later, a colleague remarked to her that those three words—"River of Grass"—"changed everybody's knowledge and educated the world as to what the Everglades meant."[3]

Conducting the research for and then writing the book absorbed Douglas for five years. More than a scholarly treatise on the natural history of the Everglades, the book includes a human history of Florida (which had not been documented) and discussions of the state's geography, weather patterns, geology, archaeology, and anthropology. Rinehart released the book in November 1947. By Christmas, all seventy-five hundred copies of the first printing had been sold. Since then it has continued to sell at the rate of ten thousand copies a year, engendering worldwide interest in the Everglades, which are unique on the planet.

Until the 1960s, when ecologists began to understand the essential role of wetlands in the functioning of the earth's ecosystems, people generally considered them mosquito- and snake-infested swamps good for nothing until they could be drained. As early as 1845, the year that Florida became a state, the legislature urged Congress to survey the Everglades with a view to "reclaiming" (draining) them.

**MARJORY STONEMAN DOUGLAS WITH GEORGE BILLIE
IN A CANOE IN THE EVERGLADES.**

It was an idea more explosive than dynamite, which would change this lower Florida world as nothing had so changed it since the melting of the glacial ice four thousand years ago. . . . The drainage of the Everglades would be a Great Thing. Americans did Great Things. Therefore Americans would drain the Everglades. Beyond that—to the intricate and subtle relation of soil, of fresh water and evaporation, and of runoff and salt intrusion, and all the consequences of disturbing the fine balance nature had set up in the past four thousand years—no one knew enough to look. They saw the Everglades no longer as a vast expanse of saw grass and water, but as a dream, a mirage of riches that many men would follow to their ruin.[4]

In 1850 Congress passed the Swamp Lands Act enabling states to "reclaim" wetlands by means of levees and drainage. Floridians believed the Everglades could be drained and put into cultivation at small cost. In 1905 Governor Napoleon Bonapart Broward swore that draining the entire Everglades could be accomplished for one dollar per acre. Marjory Stoneman Douglas's father, founder and editor of the *Miami Herald*, denounced the notion in his editorials, though he had no scientific information to back up his intuitions. Douglas said the governor and hosts of other persons hated her father for his stand.

But in the name of flood control and creating farmland for sugarcane growers and cattle ranchers—though without any scientific evaluation of the environmental consequences—the U.S. Army Corps of Engineers began ditching, diverting, channelizing, and diking. "The engineers wanted to 'correct' the flood plain system, as opposed to understanding it. . . . They dug canals to drain more water off the land. They did this to please the big agricultural interests around the lake [Okeechobee]."[5] Before they were finished, the engineers had created fourteen hundred miles of levees and canals, plus eighteen pumping stations—the largest water project in the United States.

At the same time, the state and federal governments endeavored to "get rid" of American Indians in Florida by bribing them to move out of the state. But many of them moved deep into the Everglades, where they remain today.

In the early 1940s a devastating fire swept through the former wetlands following a long drought. It shocked and awakened people to what drainage had done. And by the 1960s Floridians suspected they were ruining one of the largest ecosystems of its kind. The Everglades were dying. By the 1980s populations of wading birds had plummeted by 90 percent. Runoff of chemicals and fertilizers from sugarcane fields and livestock operations had polluted the ecosystem. Freshwater supplies dwindled even as the human population boomed. Sea water had crept into well fields.

Meanwhile, Douglas had been pursuing her writing career. After arriving in Miami in 1915, she became a journalist—at the time the only woman reporter—with the *Miami Herald*, interrupting that first career to serve in the Red Cross in France during World War I. After the war, as an associate editor with her own column, she often wrote about the Everglades, as well as Florida's landscape and geography. After three years as an associate editor, she quit the high-pressure newspaper job. For the next fifteen years she reveled in the independence of being a freelance short-story writer, selling to the *Saturday Evening Post* and other popular magazines. She also tried novels and wrote one-act plays for Miami's theater group, on whose board she served.

Eventually tiring of writing fiction, Douglas turned to nonfiction books and articles. In the 1930s she told the editor of the *Saturday Evening Post* about the trouble conservationists were having with poachers in the Everglades. Himself an environmentalist interested in Florida's birds and the wardens the National Audubon Society had hired to guard them, he assigned her to write an article about the poaching. Later, the success of *The Everglades*, her first book-length nonfiction work, led her to write six other books about Florida.

❧ The descendant of a seafaring Cornishman on one side and Quaker "rabble-rousers" on the other (one of whom had been a leader of the Underground Railroad), Douglas grew up in Taunton, Massachusetts. When Marjory was six years old her mentally ill mother abandoned her father and returned with Marjory to her parents' home. For fifteen years she did not see her father, who had relocated to Miami. In the meantime, she earned a bachelor's degree in English composition from Wellesley College. Literary magazines published her early college essays. The elocution course she took led to her being elected class orator and later enhanced her effectiveness as an activist.

The same year Douglas graduated from college, 1912, her mother died of cancer. Then, after a two-year marriage to a man thirty years her senior (who turned out to be a check forger and an alcoholic, and ended up in prison), she went to Miami in 1915 to obtain a divorce and reunite with her father. She decided she was not the marrying kind anyway, being more interested in books and writing. She also was a skeptic and a dissenter, having been a suffragist, a feminist, and a civil rights activist.

❧ In 1927 the need to protect the Everglades led to the formation of a citizens' committee—including Douglas—to make them part of the national parks system.[6] Until this year, Douglas was the committee's only survivor. The Florida Federation of Women's Clubs purchased the first parcel of land for the park—a hardwood hammock threatened by agricultural development. The federation donated it to the state. Douglas and the park committee spent twenty years educating the public and lobbying elected officials before Everglades National Park became a reality. By then the natural system had been significantly diminished. The park designation came in 1947, the same year that Rinehart published Douglas's book *The Everglades*. The park has since been called the crown jewel of the national system.

Twenty more years passed, however, before the threat of an oil refinery on the shores of lower Biscayne Bay and plans for a jetport in the middle of the Everglades galvanized Douglas into becoming a full-time environmental activist. "The Everglades were always a topic, but now they promised to become more than that. They promised to become a reason for things, a central force in my existence at the beginning of my 80th year. Perhaps it had taken me that long to figure out exactly what I was able to contribute, and for me to marshall my forces."[7] She was an environmentalist long before the word existed, but her activism bloomed in 1969 when she organized and became the first president of a citizens' action group, Friends of the Everglades.

The natural southward flow of water in the ecosystem had already been cut off by a cross-state highway, the Tamiami Trail. In addition, a levee along the southern edge of Lake Okeechobee and various canals had diverted water away from the Everglades. Even though her sight had begun to fade, Douglas lectured any groups willing to listen, recruiting members for Friends of the Everglades. She issued press releases and lobbied elected officials. Under her leadership, Floridians succeeded in defeating both the refinery and the jetport. The Friends of the Everglades, working with the national park staff, also succeeded in having culverts installed under the Tamiami Trail and some old canals filled in, thus increasing the water flow into the park.

"Since 1972, I've been going around making speeches on the Everglades all over the place. No matter how poor my eyes are I can still talk. I'll talk about the Everglades at the drop of a hat. Whoever wants me to talk, I'll come over and tell them about the necessity of preserving the Everglades. Sometimes, I tell them more than they wanted to know."[8]

In the 1960s, however, the Everglades ecosystem suffered a crippling blow when the Corps of Engineers straightened the Kissimmee River's natural meandering route as it snaked through a three-mile-wide flood-plain. Converting the river into a straight canal took ten years and cost $32 million. By the time the project was completed, the burgeoning environmental movement forced the Corps of Engineers to admit that straightening the Kissimmee was an ecological mistake. Wildlife disappeared. The river's natural cleansing function ceased. Environmentalists called the channel "the Ditch."

Of the almost three million original acres of the Everglades, half has been destroyed since 1900. But in 1985, the Everglades Coalition, comprising twenty-five nonprofit environmental groups, began collaborating with state and federal government agencies and the Corps of Engineers to develop a plan for restoration. With the ecosystem on the brink of biological collapse, work commenced in the late 1990s to implement the plan, re-creating the Kissimmee River's original oxbowed configuration as much as possible. Un-building the canal has been called an eco-correction of unprecedented magnitude. But it will revive wetlands, bring back wildlife, and improve drinking-water quality. The project will take ten to fifteen years and cost three to five billion dollars.[9] The Everglades restoration project is being hailed as a world-class event, as the Kissimmee will be the first river in the world to be restored.

The overall Everglades restoration plan also calls for about a fifth of the sugarcane farmland to be purchased by the federal government and taken out of production. Meanwhile, in November 1996, Florida's citizens approved a constitutional amendment mandating that polluting agribusinesses should be held primarily responsible for environmental cleanup costs; implementing the details was left to the Supreme Court of Florida. The Everglades have, however, been designated as a Wetland of International Significance, an International Biosphere Preserve, and a National Wilderness Area.

The invasion of foreign plants and trees poses a more recent threat. Melaleuca trees from Australia, pepper plants from Brazil used as ornamentals around homes, and an old-world climbing fern that escaped from a nursery have turned half a million acres of wet grassland into forest and thicket. The fast-reproducing melaleuca, which has no natural enemies in the United States, takes over the land and kills the saw grass, which provides habitat for alligators and other wildlife. Until a biological control can be tested, destruction and containment of the exotic species have been undertaken using heavy machines. Invader plants from foreign countries constitute the second-greatest threat to endangered species all over the United States.[10]

Marjory Stoneman Douglas's environmental causes—in addition to working for the restoration of Lake Okeechobee and the Kissimmee River and campaigning for wetlands, birds, and mangroves—encompassed the plight of the endangered Florida panther and wood stork, preserving the coral reefs of Key Largo, and abolishing hunting in Big Cypress Swamp, which she believed deprives the few remaining Florida panthers of their prey. Over the years she rescued the Everglades from plume hunters and from developers wanting to erect bridges and build suburbs and an airport. Her Friends of the Everglades group now boasts more than six thousand members; it conducts environmental education programs, disseminates information on the Everglades, and trains citizens to be activists. Her awards include the Presidential Medal of Freedom, conferred by President Clinton in 1993. On her ninety-fifth birthday, Florida named its new Department of Natural Resources building in Tallahassee after her.

Douglas gave this advice to activists:

> Join a local environmental society, but see to it that it does not waste time on superficial purposes. . . . Don't think it is enough to attend meetings and sit there like a lump. . . . It is better to address envelopes than to attend foolish meetings. It is better to study than act too quickly; but it is best to be ready to act intelligently when the appropriate opportunity arises. . . . Speak up. Learn to talk clearly and forcefully in public. Speak simply and not too long at a time, without over-emotion, always from sound preparation and knowledge. Be a nuisance where it counts, but don't be a bore at any time. . . . Do your part to inform and stimulate the public to join your action. . . . Be depressed, discouraged and disappointed at failure and the disheartening effects of ignorance, greed, corruption and bad politics—but never give up.[11]

Douglas died in May 1998 at the age of 108.

Douglas's lyrical descriptions in *The Everglades* inspire people to value these wetlands simply for their beauty, for their right to just *be*, rather than for their value to humans. Fifty years ago, in a chapter entitled "The Eleventh Hour," she wrote words that still apply today:

> There is a balance in man also, one which has set against his greed and his inertia and his foolishness; his courage, his will, his ability slowly and painfully to learn, and to work together.
>
> Perhaps even in this last hour, in a new relation of usefulness and beauty, the vast, magnificent, subtle and unique region of the Everglades may not be utterly lost.[12]

MARJORIE CARR (1915–1997)
Saves a Wilderness River

Halting a U.S. Army Corps of Engineers project already under way was unheard of in the 1960s. But Marjorie Carr—another Florida environmental matriarch of the twentieth century—intended to do just that, taking on not only the engineers but also the state and federal bureaucracies.

At issue was the fate of a canopied subtropical river that the Indians had called "Ockli-Waha," or Great River, for them an important travel and hunting resource. Fed by lakes in north central Florida, the Ocklawaha flows north alongside the Ocala National Forest for sixty miles through a mile-wide valley it has carved out over the centuries, joining the St. Johns River near Palatka south of Jacksonville. The Ocklawaha is the St. Johns's largest tributary. In a 1965 article, Carr remarks: "Throughout its course, the river twists and doubles back and forth in a well-defined, heavily forested valley so that its actual length is a third again as long as its valley . . . a hydric hammock or tree swamp rich with fauna. . . . The valley forest serves as a safe highway and sanctuary for wildlife over an enormous area."[13] In addition to its ecological value, it provided recreation for campers, hikers, canoeists, and fishermen.

Under an authorization bill enacted by Congress during World War II, the U.S. Army Corps of Engineers intended to gouge a canal across northern Florida diagonally from Palatka to a settlement called Yankee-town on the Gulf of Mexico. The plans for the Cross-Florida Barge Canal called for using existing waterways in part, damming the Ocklawaha River, creating reservoirs, and dredging a channel to accommodate major shippers' barges that would carry pulpwood, fertilizer, petroleum, and chemicals. The canal would allow shippers to bypass southern Florida. One sympathetic politician forecast that it would make central Florida a major industrial center like the Ruhr Valley—horrifying Floridians who preferred forests and tourists.

In 1962 Carr, a Phi Beta Kappa biologist and zoologist with a gradu-ate degree from Florida State University, began investigating the ecologi-cal consequences of the proposed canal; in the process she accumulated suitcases full of documents and maps concerning the Ocklawaha River. Three years later, she began an intense lobbying campaign to save the river and have the canal deauthorized, pointing out to decision makers that, since World War II, values had changed and wilderness areas were disappearing faster than anyone could have predicted. Working at first within her local Alachua Audubon Society, she wrote hundreds of letters, appealing to state legislators, U.S. senators and representatives, and even to the president's wife, Lady Bird Johnson.

A skilled problem solver, Carr recruited a cadre of natural scientists, lawyers, and economists who worked with her to do environmental and cost-benefit studies. They also worked to develop a proposal for an al-ternate route for a cross-state barge canal—if one had to be built—preserving the Ocklawaha and saving the Corps of Engineers over sixteen miles of digging. Economists predicted that the canal would not be used as much as promoters thought and would be less valuable in the future, benefiting only a few large shipping companies at the expense of Florida's taxpayers. The conservationists argued that an unspoiled wild and scenic river might prove more valuable for eco-tourism revenue than a canal.

In 1965 this small group, led by Carr, posted letters to all Florida conservationists warning them that the canal, as then planned, would obliterate forty-five miles of the Ocklawaha River, submerge 27,350 acres of riverine forest, and destroy one of the region's few remaining wilder-ness areas, one critical to native plants and animals. Furthermore, it would pose a pollution threat to surface water and to the drinking-water aquifer.

Nevertheless, the Corps of Engineers—claiming it did not have the authority to change the route (which proved to be false)—proceeded with its plans. In 1968, as the first step in the project, it erected Rodman Dam on the Ocklawaha, choking a sixteen-mile section of the free-flowing river, fragmenting the wildlife habitat, and converting nine thousand acres of rich forestland to a shallow, soupy, weed-filled reservoir. Maintaining the dam and reservoir would cost Florida taxpayers a million dollars a year.

Many people would have given up. But in 1969, Carr marshaled her scientist and lawyer colleagues and other experts and, determined to publicize the undisclosed facts about the negative environmental impacts of the Cross-Florida Barge Canal, formed a new nonprofit citizens' group—Florida Defenders of the Environment (FDE). They adopted as their motto FDE Gets the Facts. Taking other experts with her, she besieged members of Congress, state representatives, and agency heads, educating them with the facts the FDE had uncovered. She pioneered innovative training seminars for grassroots activists, teaching them negotiating skills.

In 1970 the FDE bombarded the media with the results of its scientific and economic studies. These confirmed that the canal would indeed threaten surface- and groundwater quality because of potential oil and chemical spills and other pollution from the barges. Carr persevered against formidable bureaucratic and special interest opposition until officials awoke to the canal's adverse consequences. Many species were already disappearing, including striped bass, mullet, catfish, and eels; manatees were being crushed in the dam structures. Particulate nutrients became trapped in the reservoir, blocked by the dam from flushing into the St. Johns River.

The following year, work on the barge canal came to a halt when the FDE, with help from attorneys at the Environmental Defense Fund (EDF), won an injunction in the U.S. District Court ordering the engineers to suspend construction. And on the recommendation of the President's Council on Environmental Quality, which stated that the project could jeopardize the area's unique wildlife and destroy a region of exceptional beauty, the president issued an executive order terminating the project to forestall a "serious environmental catastrophe."

Not satisfied with having defeated the Corps of Engineers and having rescued from destruction the last twenty-mile section of unspoiled river and floodplain, Carr spent the next two decades working to have the project officially deauthorized in Congress and to have the federal land that had been set aside for the canal turned over to the State of Florida as a conservation area.

The canal remained an authorized federal project, however, until November 1990, when President George Bush signed a bill Congress had passed in 1986 deauthorizing it. The Florida legislature had enacted a related bill. And late in 1992 the federal government turned over the land to Florida. In a letter inviting Carr to the transfer ceremony, the governor cited her as the person most responsible for bringing about the transfer. Seizing the opportunity to create a world-class eco-tourism destination while preserving the environment, Carr and the FDE collaborated for months with state agencies in developing a management plan for the conservation area, which would be part of a north Florida network. The plan, endorsed by the state legislature, featured hiking, camping, fishing, bicycling, and canoeing facilities. Today the Cross-Florida Greenway is a model for recreational trail systems elsewhere.

But, convinced that riverine systems must be preserved as a functioning whole in order to work properly, Carr and the FDE had set yet another, even higher, goal. They insisted that every vestige of the ill-advised canal be erased. That meant draining the reservoir, tearing down Rodman Dam, and restoring the floodplain forest. Researchers discovered that the Ocklawaha River's channel still existed at the bottom of the reservoir and would readjust when the reservoir was drained. Twenty springs could then re-establish themselves.

In the final years of the twentieth century, after decades of draining, dredging, filling, clearing, and paving, the age of restoration had come to Florida. The economic and environmental arguments for restoring the Ocklawaha proved to be overwhelming. Within a few years, savings from halting the operation of the dam and reservoir would pay for the restoration. And in the long term the Ocklawaha would maintain itself at no cost.

Although at the time the land was turned over to Florida the governor and his cabinet had voted unanimously to restore the river—a plan also endorsed by the state's Department of Environmental Regulations and its Fish and Game Commission—a few state senators, representing local fishing clubs, opposed draining the reservoir. Finally, in 1995 the Florida legislature defeated attempts to keep the dam and reservoir; the several agencies involved with the restoration work began preparing permit applications. As of early 1998 the permit process grinds on, with only one state senator threatening to challenge the permits so as to stall draining the reservoir and demolishing Rodman Dam.

In Carr's view, "Now is the time for government to atone for its past shameful weakness that resulted in a twenty-five-year-long bondage for the Ocklawaha, millions of tax dollars wasted and disillusionment with government by the public."[14]

Carr—married for fifty years to the sea turtle authority Archie Carr and the mother of five children—served as founding president of Florida Defenders of the Environment for sixteen years and remained on its board until her death in 1997. She earned a reputation among businesses, citizens' groups, and government agencies for her ability to promote understanding, engender scientifically informed decisions, and motivate change. For example, she served on an environmental advisory panel for a phosphate-mining company, working with its executives to develop mine reclamation plans. Although protecting the Ocklawaha River and its wilderness area always occupied top priority for Carr, she also exercised influence on other environmental issues in Florida, among them the Everglades and Kissimmee River restoration, preserving the Suwannee River Basin ecosystem, and protecting the endangered Florida panther and sea turtle nesting areas.

Carr observed in the April 1990 issue of FDE's newsletter, the *Monitor*:

> There is a feeling abroad in the land that, if nurtured, could bring about the saving of the world. It is the growing perception that it is ethically right to care for the earth. For years we have laboriously hunted for economic justification for every act of conservation. What good is a snail darter? What is the value of a lousewort?

But an environmental ethos is evolving. We are maturing to the point where we can simply say, "This is my home place. I like it. Don't mess it up."[15]

CHRISTINE JEAN (1957–)
Defends the Loire

Across the Atlantic, Christine Jean juggled mothering two small children with rescuing the Loire, France's longest river and Europe's last wild riverine system.

In 1980 a syndicate of politicians, headed by the mayor of Tours, hatched a plan to "improve the Loire" by erecting four dams and dikes in the river's basin. The project's promoters offered the usual arguments for dams—irrigation and flood control.

The dams would have annihilated the river valley, drowning islands that are essential wildlife habitat, wiping out traditional villages and scenic gorges (which shelter ninety-seven protected animal species), inundating pasturelands and sandbanks, and disrupting the ecological balance between the floodplain and the river. Moreover, the main dam would have further obstructed Europe's longest migration route for Atlantic salmon. The trip inland originally took only three months, but since the damming of the Loire tributaries the trip takes a year and the salmon spawn only in the main tributary.

Floods actually enhance watersheds by restoring soil fertility and replenishing underground aquifers. But rivers require room for flooding to occur. Rather than controlling floods, dams make them worse—merely moving the flood damage from one location to another. The mistake, environmentalists contend, has been allowing persons and businesses to build on floodplains.

The Loire Valley, which the French refer to as the country's garden, is where wine, dairy products, vegetables, fruit, and meat are produced; it is a vacation haven as well. Migratory birds feed and rest on the floodplain, and many species live there throughout the year. The Loire rises from springs in the Cévennes Mountains near Le Puy in south central France, flows northwest to Orléans, and then southwest past Nantes

(Christine Jean's home city), where it empties into an estuary and the Atlantic Ocean. The river—whose flow can vary from a trickle to a torrent—drains more than a fifth of France.

Passionate about the river, Jean refused to witness its demise. As an ecologist, she viewed the proposed project as reflecting an egotistical mayor's obsession with wanting to create a "great work." She felt that greedy construction companies seeking profits from development opportunities fueled his mania.

Her heavy responsibilities as a young mother notwithstanding, Jean tackled the challenge of bringing together French citizens to oppose the dams. The national government had been slow to recognize the need to protect the environment, creating a climate not conducive to environmental activism. In fact, the French government later admitted responsibility for the infamous 1985 bombing of the Greenpeace ship *Rainbow Warrior* to forestall interference with nuclear testing in the South Pacific.

Nevertheless, a number of small regional and local environmental groups shared Jean's commitment to saving the Loire. In 1986 she began working for the World Wildlife Fund to combine the groups into a nationwide coalition called Loire Vivante (Living Loire). She said it was the first time that so many organizations had agreed on an issue.

With initial funding from the World Wildlife Fund for Nature/France/Germany/International, Jean and her colleagues at Loire Vivante inaugurated a program to educate the public, engage the media, sponsor conferences, stage demonstrations, and ultimately file a legal suit against the dam-building consortium. Christine Jean, trained in agronomy and holding a master's degree in ecology, drew on her scientific expertise to help lobby decision makers, counter engineers' arguments, and inform citizens. She and her associates pointed out that the era of dams had passed. In Germany and the Netherlands dams were being torn down and rivers rehabilitated. It did not make sense, she said, to carry out a project that ultimately must be undone. Rather, forbidding people to build on floodplains was the best protection against flood damage.

In 1988 environmentalists in the Le Puy region, the site selected for the largest dam, organized a coalition affiliate called S.O.S. Loire Vivante to coordinate activities against the upstream basin's dam. S.O.S. Loire Vivante soon grew to be the largest of the affiliates.

As an initial step in the overall campaign, Loire Vivante's NGO network garnered thirteen thousand signatures on an anti-dam petition addressed to President François Mitterand. At the same time they demanded that a comprehensive environmental impact assessment of the plan be conducted in advance of any construction, encompassing the effects on the entire region. To guarantee that work on the dam could not begin, Loire Vivante bought a piece of land in Le Puy near the site chosen for the first and largest dam, the Serre de la Fare dam. In February 1989 a small group of activists established a round-the-clock campsite, prepared to place their bodies, if necessary, in the way of bulldozers. Persons of all ages became campers. On weekends up to a thousand supporters came to Le Puy to join the show of protest, leading to broad media coverage and focusing international attention on France's environmental problems.

Meanwhile, Jean stumped the country, speaking and gathering support for the cause. Several months later, she and her network organized an anti-dam rally in Le Puy. Ten thousand persons from all over Europe attended, waving flags and banners. Jean described having thousands of demonstrators walking behind her down the street as an awesome experience. The campaign to save the Loire River proved to be the most talked about environmental issue in two decades.

In parallel with protesting the dams, Jean and Loire Vivante worked on developing an alternative proposal for river management, especially flood control. It emphasized river restoration, biodiversity protection, and re-establishment of the Atlantic salmon population, which was near extinction.

While all this was going on, Jean's husband, a city planner, was reassigned for a time to Montpellier near the southern coast of France, over four hundred miles away; this meant the family could get together only about once a month. At that point, Jean found mixing her private and professional lives difficult. Caring for the children alone was not easy. But she relished the challenge of fighting for the Loire too much to give it up.

In 1991 Loire Vivante won a lawsuit against the dam syndicate. This victory led to Christine Jean and Loire Vivante receiving the Goldman Foundation's Environmental Prize the following year. But the activist campers maintained their vigil at Le Puy until 1994, when all plans for the Serre de la Fare dam were officially abandoned. Of the other three dams in the original plan, one has been built, another simply postponed, and plans resurrected "for the useless dam at Chambonchard after being abandoned in 1991."

Since cancellation of the Serre de la Fare project, the French government has launched a ten-year program, Plan Loire Grandeur Nature, embodying the principles of river management that had been developed by Loire Vivante NGOs. The plan—designed to reconcile environmental protection, population safety, and economic development from a sustainability and ecosystem preservation perspective—includes tearing down two older dams on a tributary in the upper reaches of the basin that had been blocking the salmon's migration.

The success of a natural flood control alternative to the Serre de la Fare dam as proposed by Christine Jean and the affiliates of Loire Vivante has subsequently been demonstrated during a flood on the upper Loire near the city of Brives-Charensac, where a flood caused eight deaths in 1980. Implementing the alternative plan has given the river room to flow by deepening and broadening its bed instead of diking, by clearing the riverbanks and doing new landscaping, and by moving buildings off the floodplain. A computerized radar system provides early flood-warning alerts. Loire Vivante's counterproposals have proven efficient in practice, at a third of the dam's cost.

However, serious threats to the Loire system still loom, Jean says. In 1994 a project went forward to enlarge the Nantes-St.-Nazaire port at the river's mouth to accommodate construction of a nuclear power plant in the estuary.[16]

Jean says the most difficult obstacles she encounters in her work are the perpetual lack of funds to support Loire Vivante's anti-dam campaign and the general attitude that "rivers are simply water pipes for human needs." What keeps her going in the face of hostility and setbacks? "The rewarding relationships among the people of Loire Vivante, and the positive events that happen regularly."

In partnership with the WWF, the European Rivers Network, and the International Rivers Network, the Loire Vivante coalition continues its work of nurturing a culture of conservation among citizens and monitoring river protection in France. Jean remains a member of Loire Vivante's coordinating group. She urges people to support the environmental movement by joining an organization and donating money. Her effectiveness as an eco-hero has earned her the nickname Madame Loire. The growing popularity of France's Green Party testifies to Jean's influence on public opinion. In an election in the early 1990s, the party won 22 percent of the votes in the region that was to have been dammed.

To become aware of other creatures as individuals is to discover that life is a mansion with many rooms.

LORRAINE ANDERSON
Sisters of the Earth

Women and Wildlife

HARRIET HEMENWAY (1858–1960)
Outraged Society Lady

Without any feeling of guilt, Harriet Hemenway had been wearing feathers—even whole birds—on her hats for years. All stylish women in the Boston social register did. Then one winter day in 1896 she read a description of what a Florida heron rookery looked like after being raided by plume hunters. Dead birds lay in the mud. The skins where the long plumes grew on the backs of the adult birds had been ripped off and their bodies left to rot. Orphaned young clamored for food. Others lay dead in the nest, heads dangling over the edge.

Aghast, Hemenway, who came from a family active in public service and possessing wealth and social clout, reached for her copy of the *Boston Blue Book* and strode across the street to the home of her cousin, Minna Hall. Together they marked the names of women they thought might join a group to protect the birds. Aware, however, that such a group had to reach people outside their own circle, the two women convened a meeting in Hemenway's home of well-known ornithologists and society leaders to talk about the problem of hunting plumed birds. On the spot the group voted to create the Massachusetts Audubon Society, whose purpose was to discourage the purchase and wearing of feathered hats and to protect wild bird species.

With the organization in place, Hemenway set in motion the next part of her strategy—a series of tea parties. Over refreshments, she entreated her guests to give up wearing feathers and to join the Audubon Society. By the end of a year, the new society had gained nine hundred adherents. The roster of founding members included the names of prominent and affluent Massachusetts families. At the same time, Hemenway and the fledgling Audubon Society waged a campaign to persuade milliners to design featherless hats, to make wearing them fashionable, and to lobby for legislation outlawing the slaughter of birds.

During the business boom after the Civil War, plume hunting had flourished. In New York, plumes considered to be the most elegant—the long egret nuptial feathers known by the French term *aigrettes*—brought seventy-five cents each. To reap twenty or thirty plumes, hunters would wipe out an entire colony. One Florida hunter estimated that in only one season he had killed 130,000 birds. All birds, even crows, were at risk. In 1886, a decade before Hemenway launched her crusade, a New York City ornithologist walking on Fourteenth Street identified feathers from forty species on hats he passed, including common terns, grebes, bobwhites, blue jays, woodpeckers, robins, warblers, cedar waxwings, and a scissortailed flycatcher. By the end of the nineteenth century, at least five million birds a year were being killed for their feathers.[1]

Within two years after the founding of the Massachusetts group, women in fifteen other states had formed Audubon Societies. In 1905 some of them consolidated into a National Association of Audubon Societies (later renamed the National Audubon Society). Some state groups, including Massachusetts Audubon, elected to remain independent of the national organization. Gradually, under pressure from the various societies, state after state began tightening its conservation statutes. In the New York legislature, a dedicated young birder, Franklin D. Roosevelt, backed by a deluge of petitions from women's clubs, succeeded in winning the day over milliners arguing that a restrictive bill would eliminate twenty thousand jobs. But not until 1913 did the U.S. Congress enact national legislation to protect migratory birds. Already Hungary, in initiating the first International Ornithological Congress in 1890, had focused world attention on the destruction of insectivorous migratory birds in Europe, which was threatening food crops there.

Hemenway worked with the Massachusetts Audubon Society mostly behind the scenes, providing financial support and advice, but she often entertained as many as sixty members at a time in her home to recharge their energies. And sometimes she stepped in when squabbles among the staff required arbitration. In those early days, women in Massachusetts Audubon held positions as officers in equal numbers with men. And women headed 114 of the 118 local chapters established in the state.[2]

Hemenway proved to be a formidable advocate for feathered species. Her grandfather had accrued a fortune in textile manufacturing. Her father, an antislavery agitator, had founded colleges and a town. Her husband, Augustus Hemenway, came from an equally wealthy and public-spirited background.

Boston Brahmins had become used to Harriet Hemenway's independent nature. Outspoken and unafraid to break with bigoted social attitudes, she once spotlighted racial prejudice by inviting Booker T. Washington into her home when Boston hotels turned him away. And, unconcerned with the dictates of fashion, she wore white sneakers on her bird walks long before birders adopted them as their symbol.

In 1946 Hemenway and her cousin Minna—both then in their eighties—got together once again in Hemenway's parlor to mark the half-century anniversary of their historic huddle, which inaugurated the culture of conservation in the United States. Of all the nonhunting outdoor activities, birding has now become the most popular. Today, killing non-game birds is not only illegal but it is looked on as immoral. Birders, with their extensive experience in the field, have formed the talent pool from which three-fourths of the leaders of NGOs comprising the core of the environmental movement have come.[3]

In February 1996, Massachusetts Audubon—the most successful of the independent state Audubon Societies, with a membership of fifty-five thousand, a ten-million-dollar budget, and a generous endowment—celebrated its centennial and toasted its founder at a party in Boston's statehouse.

Harriet Hemenway, who has been referred to as bird-watchers' matron saint, died in 1960, aged 102.

MARGARET MURIE (1902–)
Watches Over the Alaskan Wilderness

America's last wild lands lie in Alaska.[4] They are federal lands owned by all Americans. Trying to keep them intact as a heritage for future generations has occasioned bitter controversy, especially during the past fifty years.

Saving some distinct areas of Alaska and their wildlife habitat began over a hundred years ago, shortly after the United States purchased the land from Russia in 1867. Executive orders in the late 1800s and early 1900s created some refuges and parks, including Alaska's first national park—the area surrounding Mount McKinley, the highest point in North America (now called Denali National Park). After World War II, tensions escalated between oil and gas interests, developers, and Alaska politicians—all of whom opposed wilderness protection—on the one hand and environmental groups on the other. The effort to save some of the state marked the environmental movement's coming of age as a heretofore unmarshaled national force with the ability to exert political influence. In the late 1970s, during the peak of the battle over Alaska's wilderness areas, the Alaska Coalition grew to fifty-five nongovernmental organizations, including six labor unions.

About twenty years later, after President Bill Clinton rang Mardy Murie on her ninety-third birthday (18 August 1995) to promise her he would veto any plan to allow mineral development in the Arctic National Wildlife Refuge (ANWR), she breathed easier. Clinton is the third U.S. chief executive to recognize Murie's role in helping preserve the Alaskan wilderness. This cause has engaged her for most of the twentieth century.

Even though she had grown up in the frontier town of Fairbanks in the early 1900s, often roaming the nearby forests and swamps, Murie's first experience in Alaska's wilderness came in 1924. As a young bride, she set off on a 550-mile dogsled honeymoon into Alaska's Arctic interior with Olaus Murie, a renowned wildlife biologist and artist. They were en route to the Endicott Mountains above the Arctic Circle in the Brooks Range to conduct caribou studies for the U.S. Biological Survey (now the Fish and Wildlife Service). Once the Koyukuk River had frozen over in the early fall, they mushed up the river by dogsled. In her autobiographical memoir, *Two in the Far North*, she recalls:

> An ideal day to hit the trail: twelve below, just right for mushing. We were both running, and I was soon too warm; I threw back the parka hood and pulled off the red toque; the crisp air felt good on my bared head. How light my moccasined feet felt, padding along on snow-sprinkled ice at a dogtrot, exhilaration in every muscle responding to the joy of motion, running, running without getting out of breath.[5]

This marked the beginning of forty years of adventure and working together in the wilderness. Murie relished every moment. Their expeditions into the Arctic cumulatively amounted to six years of fieldwork— along the Old Crow, the Koyukuk, the Porcupine, and the Sheenjek Rivers. They studied and documented the topography, vegetation, birds, mammals, and habitats, categorizing them into ecosystems and collecting specimens. A wildlife filmmaker accompanied the small group of scientists on the Muries' 1956 ten-week summer expedition to the Sheenjek River Valley in what is now called the Arctic National Wildlife Refuge, where from 23 May to 20 July the sun never sets. The resulting film, *Letter from Brooks Range*, captured spectacular wilderness landscapes and proved to be an effective lobbying tool.

Of that expedition, Murie notes:

> In this day and age, it is a rare experience to be able to live in an environment wholly nature's own, where the only sounds are those of the natural world. . . . [T]here was the splash of a muskrat diving off the edge of the ice; tree sparrows and white-crowned sparrows sang continually. . . . We heard the scolding chatter of Brewer's blackbirds and, what at first seemed very strange up there in the Far North, the voice of the robin, our close friend of all the mild, domesticated places. . . .
>
> Far across from us, we sometimes heard the indescribably haunting call of the arctic loons, and then all binoculars would be snatched up for a glimpse of these beautiful patricians of the North. . . .
>
> I have watched a band of fifty caribou feeding back and forth on a

flat a quarter mile away; ptarmigan soaring and cluck-clucking . . . ; cliff
swallows hurrying by; Wilson snipe and yellowlegs calling; gray-cheeked
thrushes singing. The three young scientists are beside themselves with all
there is to see and do and record.[6]

The Muries' photos, the data they collected, and their writings
helped lay the groundwork for the establishment in 1960 of the nineteen-
million-acre Arctic Wildlife Range, as it was first called, by executive
order of Secretary of the Interior Fred Seaton. This area stretches from
the Beaufort Sea two hundred miles south to the Porcupine River and
features the highest peaks, the most extensive glaciers, and the least disturbed terrain in the region.

The Muries' easygoing lobbying among groups in Alaska and Washington, D.C.—especially in the late 1950s, during bargaining over state-
hood—triggered grassroots momentum, and the idea of an Arctic wilderness started to germinate. In recognition of her efforts, the Department
of the Interior later conferred its Citizen Conservation Award on Margaret Murie.

When Murie children came along, two boys and a girl, they joined
their parents on wilderness expeditions. When the first child, Martin, was
ten months old, they took him along in a wooden packing box (covered
by a mosquito net). Accompanied by Jess Rust, a colleague of Olaus, they
embarked on a three-month assignment banding geese and observing caribou along the Old Crow River in northeastern Alaska. Hordes of mos-
quitoes compelled them to sleep on sandbars in the middle of the river.
When their motor's crankshaft broke in the rapids, they had to continue
250 miles upriver by poling a motorless scow full of supplies that they had
been pushing. The return downriver through the rapids in a powerless,
unwieldy scow challenged their whitewater skills.

Murie found family life in the wilderness simpler than in town. For
example, when Olaus was assigned in 1927 to direct a field study to determine the cause of an elk die-off in the Wyoming Tetons, the family established a permanent home base there. They camped in a tent all summer
in a high meadow in the Teton wilderness, the elks' summer range. The
children spent their days in the open air, had healthy appetites, were
never sick or cross, and needed only the simplest clothes. Murie noted in

her memoir that in wilderness camps there was no vacuuming or dusting and no telephone to answer.

It became obvious to the Muries early on that with the threat of highways, resorts, dams, and oil and gas explorations intruding on unspoiled areas, the only way to save some wilderness heritage was for them to become involved in these controversial public issues. In 1935, with a small group of colleagues, they organized the Wilderness Society. At different times both Muries served the society in leadership roles.

Although she had worked alongside her biologist husband for forty years, keeping a journal and assisting him with his wildlife field studies, it was only after his death in 1963 that Murie emerged as an environmental activist in her own right. Once she accepted that the pain of his loss would never go away, she decided to build a new life on top of it. She began writing, giving talks to conservation groups, speaking with individuals wherever she went, analyzing environmental impact statements, testifying at congressional hearings, and lobbying for conservation causes. Her lyrical written and oral descriptions of her experiences in the wilderness won allies. She also worked for the Wilderness Society for a while, serving on its council and as vice president. In 1964, when her work bore some fruit and Congress passed the Wilderness Act, President Lyndon Johnson included Murie in the Rose Garden signing ceremony.

She made several trips back to Alaska. In 1975, when she was seventy-three, Murie was doing an aerial survey of a remote wilderness area in connection with Congressman Morris Udall's proposed Alaskan lands legislation. The small floatplane she was in—copiloted by her friend Celia Hunter, an Alaska conservation colleague, a World War II WASP, and a bush pilot—crash-landed when the motor failed. Seventeen hours later a helicopter pilot spotted the party and rescued them.

In the years following the creation of the Arctic Wildlife Range in 1960 (a step that only partially protected the area), Celia Hunter and her Alaska Conservation Society helped monitor the region. They protected it from such development schemes as a hydroelectric dam, "wolf control," a natural gas pipeline, and a plan to use atomic explosions to create a deepwater port.

The controversy over legislation to protect Alaskan lands reached the boiling point in the late 1970s, mobilizing the Alaska Coalition's vast network of grassroots activists and bringing unprecedented cooperation and sharing among NGOs. When congressional committee hearings were scheduled in several key cities around the country, hundreds of environmental advocates jammed the meeting rooms.

In early June 1977, Murie was first to speak at the Denver session. She said she was testifying as an emotional woman, that beauty itself was a resource—that Alaska should be allowed to *be* Alaska, which was its greatest economy. She hoped, she said, that our country was not so rich that it could afford to allow the Alaska wilderness to be lost, nor so poor that it could not afford to keep it. Her testimony inspired a spontaneous standing ovation.

On 21 July 1980, President Carter held a kickoff event at the White House for the Alaska Coalition's final lobbying push. Murie took a turn at the podium in the East Room to energize the group for its critical last-minute effort.

When the Alaska National Interest Lands Conservation Act finally passed Congress, President Carter shook Murie's hand at the signing ceremony, thanking her for her role in making it happen. This landmark legislation, the greatest land preservation act in the nation's history, set aside more than one hundred million acres for parks and refuges. However, in the battle over the legislation, conservationists had not been able to obtain wilderness area designation for all of the Arctic National Wildlife Refuge, leaving the coastal plain "open for further study" and vulnerable to mineral exploration.

Developers and their supporters in Congress have attempted ever since to justify oil and gas development there, invoking national interest. However, the *Exxon Valdez* oil spill in 1989 helped dampen public tolerance for the Arctic drilling idea. It was when members of Congress, influenced by the petroleum lobby and Alaska's governor, sought to include projected oil revenues from the ANWR coastal plain as part of the 1996 budget bill that President Clinton telephoned Murie to assure her he would not allow it. She cautions, though, that the goings-on in Congress need to be watched. The ANWR is still in a precarious position, she says, with some interests seeing only oil and money there.

Now in her mid-nineties and with great-grandchildren, Murie lives close to wildlife in her log house on the ranch she and Olaus acquired within Teton National Park in 1946. The Snake River runs nearby, and mountains, forests, and moose and other animals surround her. A pine marten, seldom glimpsed by humans, raised families for several years under her cabin.

Over time her ranch became a gathering place for environmental activist leaders to discuss strategies for protecting the ANWR as well as other national and local issues. For a number of summers Murie also conducted weekly nature classes in her home for students from the Teton Science School. She considers this work with young people and her contribution to the creation of the ANWR two of her most important and rewarding activities.

❧ The recipient of an honorary doctorate from the University of Alaska in her later years, Murie (in 1924) was the first woman to graduate from the university. (She briefly attended Reed and Simmons Colleges, where she majored in English.) For many years she devoted her talent for writing to keeping her journal. But chapter-at-a-time prodding from an editor friend at Knopf in the early 1960s yielded *Two in the Far North*, a love story and adventure saga about her early life, courtship, 3:00 A.M. wedding in a settlement on the Yukon River, and her Alaska wilderness years studying wildlife with Olaus. In 1997, Alaska Northwest Books reissued it in a thirty-fifth anniversary edition as an American Wilderness Classic.

She wrote two others books, *Wapiti Wilderness* (*wapiti* is the Shawnee word for "elk"), together with Olaus, about their experiences during Teton wildlife research projects, and *Island Between*, about Eskimo culture.

To continue carrying her message to the world, a Wyoming filmmaker is producing a biographical documentary, *Arctic Dance: The Mardy Murie Story*, for television and video distribution. Murie's reminiscences with her friend Terry Tempest Williams, an author and naturalist, form the backbone of the documentary, which also includes archival footage. Harrison Ford narrates the text, which was written by Williams. Paul Winter and John Denver provide the music. The film's world premiere is planned for winter of 2000 in Jackson Hole, Wyoming. Proceeds from the production will help endow a Murie Institute and Museum.

Does she have hope that humans will be able to save the planet?

"Yes," Murie says, "I believe that the dedication of environmentalists in working for positive changes and putting pressure on decision makers will do it."

SARAH JAMES (1944–)
Fights to Protect Arctic Caribou

"Should a budget trick by oil lobbyists wipe out a thousand genera-tions of Arctic culture?" read an advertisement in the *New York Times*.[7] The ad referred to the attempt by Congress to attach a provision to the 1996 federal budget bill that would have allowed oil drilling in the Arctic National Wildlife Refuge's coastal plain—the sensitive calving area of the Porcupine caribou herd. Alaska's governor had spent over a million dol-lars trying to convince Congress and the president to allow drilling in the refuge, hiring a lobbying firm in Washington, D.C., at $245 an hour. Additionally, oil companies had given $400,000 to Arctic Power, a lobby-ing consortium of Alaska businessmen, unions, and citizens.

President Clinton's threat to veto the bill squelched the budget rider.

The *Times* ad had been placed by Sarah James and the Gwich'in Steering Committee, a nonprofit American Indian environmentalists' group. As the steering committee's chief executive officer, during the past decade James has been shuttling between her home in Arctic Village, a hundred miles above the Arctic Circle in northeastern Alaska, and the Lower Forty-Eight. She estimates she has made about ten trips since 1988 to Washington, D.C., meeting with individual members of Congress and testifying at six Senate committee hearings. She has visited all but the southeastern states, speaking to groups and raising funds for the Gwich'in campaign. She has even traveled to conferences in Central and South America. For example, in 1989 she pleaded for help for her tribal nation's cause at an environmental conference—The Fate and Hope of the Earth—in Managua, Nicaragua, attended by seventeen hundred persons from seventy countries. In 1992 she participated in the NGO forum of the Earth Summit in Brazil. Much of her travel abroad relates to her work as a board member of the International Indian Treaty Council and involves defending indigenous peoples' human rights.

Until James began her pilgrimages, few people outside Alaska had heard of the Gwich'in Indians. Acting as a spokesperson, she tells her people's story, lobbying on behalf of the traditional elders of her tribe, educating the public, and winning allies. As a result, financial contributions from 258 indigenous tribes from all over the world now underwrite the Gwich'in Steering Committee's work. The committee receives additional funding from churches, foundations, and individuals.

James says Gwich'in activists are about evenly divided between women and men. "In our culture we do the same things. Women do everything men do, and men do everything women do. We live in partnership. We all need to get back to working in partnerships again."

The Gwich'in (who are also called Ashabascan, the "caribou people") are the northernmost Indian nation in North America. To them the Porcupine caribou, named after the Porcupine River, are sacred. Their lives have been intertwined with the caribou herd for twenty thousand years. Their songs, dances, legends, and spirituality derive from these animals. They think of the caribou as a gift and treat them—and all other animals—with respect. They believe their people are responsible for caring for the caribou and managing their habitat; if they do, the

SARAH JAMES AT ARCTIC VILLAGE, ALASKA.

animals will in turn nurture them. Their culture and way of life depend on the health of the Porcupine caribou herd. And the health of this herd, numbering about 150,000 animals, requires that its calving and nursing grounds, located along the ANWR's coastal plain, not be disturbed.

The herd's summer calving and nursing activities need the delicate balance of climate, food supplies, and absence of biting insects and predators that characterizes the ANWR's coastal plain. Sarah James says her people view the calving grounds as a holy and sacred place where life begins, a place to be kept safe and quiet, without disruption. The Gwich'in never enter the calving grounds, James says, not even in hard times.

Each summer the herd migrates from Canada to the coastal plain, where new willow and cotton grass growth offers the high protein concentration needed by caribou cows right after calving—the period of highest energy demands. Biological studies show that when cows fail to gain weight over the summer 80 percent of their calves die.[8] Although oil companies contend that development would not disturb the caribou, Sarah James and the Gwich'in people say this is not true. Biologists agree, cautioning against introducing activity to the calving grounds. Their studies confirm a link between the level of disturbance and health problems, calf mortality, and pregnancy failures. Caribou living near the Prudhoe Bay oil fields, for example, have far fewer calves.

The seventeen Gwich'in villages, comprising about seven thousand people, are strategically situated along the caribou's seven-hundred-mile migration route. The Gwich'in wait for the animals to pass near their villages each spring and fall—times for festive celebrations—before taking only the minimum number of animals they need for subsistence. The Gwich'in believe that oil development will destroy their culture by polluting the coastal plain and destroying the caribou herd. They consider the possibility of oil drilling to be a death threat and believe it would result in their destruction. "It is a violation of human rights to destroy our culture with oil-drilling activities," James asserts, citing the International Covenant on Economic, Social, and Cultural Rights.

At a historic gathering of tribal chiefs in 1988, the Gwich'in people organized the Gwich'in Steering Committee. They also adopted a resolution protesting coastal plain development, quoting a phrase from the International Covenant—"In no case may a people be deprived of its own means of subsistence." The steering committee's resolution calls for officially designating the coastal plain of the ANWR as wilderness; this would involve an area of only 1.5 million of the refuge's 19 million acres.

James says conveying the human rights aspect of the issue to the public is one of the most difficult obstacles she encounters. "It is human rights versus oil," she says. "And it is a corporate tactic to make Native people disagree and fight among themselves." James also reminds audiences of the Memorandum of Understanding signed in 1987 by Canada and the United States in which each government agreed to protect the caribou's summer calving area, which extends into Canada's Yukon Territory. "Another difficult thing," she says, "is having to go back home to my people with bad news about what is happening in Washington and try to explain it to them in their Native language and make them understand what they need to do."

The Arctic National Wildlife Refuge, in addition to its calving area for the caribou herd, also provides habitat for a host of large mammals, among them polar bears, grizzlies, brown bears, wolves, Dall sheep, musk oxen, and moose.[9] It is, as well, the summer home of many species of birds and marine mammals. A recent analysis by the U.S. Fish and Wildlife Service points out that the ANWR is the only conservation area in the country that includes a complete range of Arctic ecosystems, functioning in balance to sustain wildlife populations. Ecologists consider it a pristine ecosystem of world significance.

Sarah James's home is in a tidy bush community of log houses called Arctic Village, located in the northernmost tip of the 1.8-million-acre Venetie Indian Reservation, just outside the Arctic National Wildlife Refuge. Tucked into the south side of the Brooks Range, Arctic Village, with a population of 150 persons, is accessible only by airplane. Large satellite dishes provide the community's main link to the outside world. Two small stores stock a few staples. The local school has a computer room accessible to James and to the community at large.

The Gwich'in people still pursue their traditional way of life, living off the land rather than working for paychecks. They spend their days hunting for meat, gathering other wild foods, creating snowshoes and tools from caribou parts, and making clothing and boots from caribou skins. Sarah James laments, "I haven't been able to live the kind of life I'm trying to protect because of all the traveling around to educate people about the issue."

James says she was born with a concern for the environment. "It's a Native American feeling about the earth. The Creator makes us be that way—we're stewards of the Earth." Although she comes from a fairly large family, only three of her siblings (two brothers and a sister) survive. At thirteen she went away to a boarding school in Oregon. "It was a struggle because of culture shock and it took me six years to finish high school." Even though her twenty-something son takes pride in his mother's efforts, he chooses not to be involved to the extent she is. However, all Gwich'in people—through their daily activities—reaffirm their commitment to preserving the ANWR wilderness and its plant and animal life.

The Gwich'in people do not oppose all mineral development in Alaska. Like all residents, they receive oil dividend checks each fall, and oil money funds their schools. However, although they admit that drilling along the ANWR's coast would produce some oil-field jobs for them and bring income for luxuries, they prefer to continue pursuing their traditional lifestyle and to preserve the land and wildlife. James believes the oil industry exaggerates the number of jobs that drilling would create. Such projections, she says, are ways of "getting their foot in the ANWR door after which they would move into other refuges." She would prefer to see "the creation of jobs restoring the earth, developing alternative energy systems, replanting trees. We need to stop taking from the earth. We've already taken enough out. We should recycle and keep going on what we've already taken."

James points out that, except for the small section of coastal plain that is part of the ANWR, the entire northern coast of Alaska has been opened for mineral exploration and drilling. There have been many oil spills, she says, and the companies have made no effort to restore the areas of depleted and abandoned oil wells.

She worries about global warming, pointing out that Alaska is already experiencing climate change.[10] It is threatening Arctic ecosystems and some of the world's most distinctive mammals—Arctic foxes, polar bears, and bowhead whales—which are unique to the region. Fifteen percent of the world's bird species breed in the Arctic.[11] "Our people will be the first to be affected," James notes. The Gwich'in are among some two hundred grassroots North American Native groups defending their homelands against environmental destruction.

What keeps her going in the face of opposition and setbacks? "I remember what the elders have always said: Do this in a good way and teach the world's people who don't know our ways. It is tiresome, but rewarding." But she is gaining media and public attention. Mention of James's fight on behalf of the ANWR wilderness and the Porcupine caribou, along with photos of her, have appeared in a number of publications, including *Business Week*, *Sierra*, and *Audubon*.

Does she have hope that we will save the planet before it is too late? "We don't have much time; but if we come together, we can make a turnaround and 'walk our talk.'"

Environmentalists all over the world agree with the Gwich'in people that the ANWR's coastal plain deserves wilderness status barring forever any development. It may happen. Bills have been introduced in Congress each year since 1985 that would declare the plain a protected wilderness area. Although the Senate bill of 1997 had eighteen cosponsors, it was stalled in the Environment and Public Works Committee. Unless or until Congress takes specific action to allow development, the coastal plain of the ANWR will continue to be managed as though it were wilderness.

SYLVIA EARLE (1936–)
Safeguards Earth's Oceans

Of her first encounter with whales during a 1977 research expedition among the Hawaiian Islands, Sylvia Earle recalls:

> Looming from deep water, well below where I floated but fast approaching, was the largest creature I had ever seen, a plump, pregnant humpback whale, now less than 30 feet away . . . then 15 . . . then a mere arm's length from collision. It was too late to worry whether or not the whale had brakes or to attempt a swift sideways slither out of her way. My fate depended entirely on the agility, perception, and attitude of a creature more than 800 times my weight. I was acutely aware at that moment that

my species has a long history of actively hunting and killing members of hers in all of the world's oceans. With that track record, there were plenty of reasons to expect hostility. There was no precedent for what I was doing, no "guidebook" with thoughtful descriptions of proper etiquette for greeting humpbacks underwater or surefire protocols to let the whales know that my intentions were entirely benign. . . .

One moment, all I could see was a massive black head, 15-foot-long flippers, and a grapefruit-sized eye, then, in a mountainous blur of motion, she swept by, miraculously avoiding contact by inches. With a small gasp, I realized I had been holding my breath . . . and so had the whale. Her great back lifted slightly out of the water as she emitted a great *whoosh* of air and salt spray, inhaled sharply, then arched back down, propelled by a modest flip of her huge tailfluke.[12]

The research team in this expedition observed humpback whale be-havior for three months, capturing on film for the first time a "singer" in action. Earle says, "Once you've met a forty-ton whale, once you've seen them as they see themselves, you will never look at them—or yourself—the same way again."

She also issues a challenge based on her encounters: "There is an urgent need for us to consider whales and other ocean life (and ourselves) in new ways that take into account the importance of the living sea as earth's life-support system."[13]

Oceans and coastal waters comprise 99 percent of Earth's dwelling places for animals and plants, contain 97 percent of the planet's water, and cover over 70 percent of its surface. Oceans govern our weather, generate the oxygen we need to breathe, and house our most extensive natural-food resource. Yet because what goes on in their depths is out of everyday sight and understanding for most people, the oceans have been treated like a big septic tank. They are the dumping ground for nuclear, industrial, and municipal wastes, sewage, hard trash, toxic agricultural chemical run-offs, and oil from spills, for example. Furthermore, over-fishing has caused the near collapse of many major commercial fisheries.

The indiscriminate wildlife carnage wrought by modern commercial fishing technologies particularly angers Sylvia Earle. Gill nets, for exam-ple, entangle such nontarget species as turtles, whales, sharks, birds, and

SYLVIA EARLE DOCUMENTING SEA LIFE, MAUI, HAWAII.

dolphins. In her book *Sea Change*, she points out: "Little is being done to change laws that permit agonizing death by suffocation, strangulation, crushing, drowning, panic, shock, slicing, spearing, or other modes of modern fishing. No one doubts that dolphins, whales, seals, and birds feel the burn of rough webbing on exquisitely sensitive skin, the slashing bite of knives and gaffs, the searing shock of separation from close-knit socie- ties . . . and no one *should* doubt that fish do as well."[14]

Public protests forced the passage of a law regulating the number of dolphins that tuna fishermen can incidentally capture and kill in nets. A death rate that ran to two hundred thousand a year in the 1970s has been reduced, although the tuna industry is still permitted to kill twenty thousand dolphins a year. Overwhaling brought several species, including the humpback whale (now protected as an endangered species), to near extinction. "If the sea is sick, we'll feel it. If it dies, we die. Our future and the state of the oceans are one. . . . [T]he ocean is the cornerstone of the systems that sustain us: every breath we take is linked to the sea."[15] Even our briny blood is evidence of our connection to the sea.

Earle, a marine botanist and biologist who earned her advanced degrees at Duke University, believes that abuse of our oceans and their wildlife stems from indifference and ignorance, even among many otherwise intelligent people. She contends this ignorance is by far the greatest threat to the sea, and sees it as related to the notion that the ocean is so resilient and so vast that there is no need to worry about what we dump in or take out. For the past quarter century she has not only pioneered exploring the oceans as a diver but has also focused on educating the public and decision makers. Through more than eighty articles in scientific as well as popular journals, three books, lecturing in more than sixty countries, participating in conferences, and appearing on television periodically, she endeavors to shape public attitudes. She also huddles regularly with businesspeople and government leaders to thrash out ecological problems. What motivates her is her conviction that along with knowing about the ocean comes caring.

Earle points out that we have barely begun exploring that part of our planet that lies under water. Although the deep sea has been mapped using electronic sounding devices, Earle says that more than 95 percent of the sea remains unexplored—never observed firsthand, sampled, touched, or understood. Much of what *has* been explored can be credited to her.

Earle's interest in the ocean began in childhood, when she lived first near the New Jersey shore and later on Florida's Gulf Coast. Borrowing a sponge diver's helmet from a friend, she made her first underwater descent at age sixteen, to take a look beneath the surface of a Florida river. Since then she has logged over six thousand hours diving in oceans all over the world. A pioneer in deep-sea research and technology, she first came to public notice in 1970, when the National Aeronautics and Space Administration (NASA) appointed her to lead a team of women scientists living underwater for two weeks in the Caribbean. She has led fifty undersea expeditions and holds the world's record for solo diving: In 1979 she spent two and a half hours in a record-breaking 1,250-foot untethered dive surveying the ocean floor and gathering information off the island of Oahu in Hawaii. Her goal is to dive to the oceans' deepest point—36,198 feet—in the Mariana Trench in the Pacific.

In 1981 Earle and a British colleague founded Deep Ocean Engineering, which designs and manufactures state-of-the-art underwater diving vehicles. Despite this business, funding for her undersea research presents an ongoing problem. Earle contends that the space program's public relations campaigns sucked funds away from ocean research. Everyone can envision walking on the moon because it is visible, she notes, whereas walking on the seafloor is more difficult to imagine.

Earle tried three times to mix marriage and career. Repeatedly she opted for her career. Though a mother of three children, she admits that being a scientist is who she is—above and beyond being a woman or wife. During their growing-up years, she even involved her children in her research projects whenever possible, engaging them as "research assistants." For example, while exploring a reef in the Bahamas, she took them along for a week of diving and interacting with dolphins.

As a leading expert on marine ecosystems, Earle's knowledge guides the ocean policies of the numerous boards and councils on which she serves. These include, among others, the Center for Marine Conservation, the World Wildlife Fund International, and the Natural Resources Defense Council. She also hosts a daily program on public radio, *Ocean Report*.

In the early 1990s she served by presidential appointment for eighteen months as chief scientist of the National Oceanic and Atmospheric Administration (NOAA), the first woman ever to hold the position. During that period she made many trips to the Persian Gulf to assess the environmental devastation inflicted by the infamous oil-well fires in Kuwait—the planet's worst oil catastrophe.

Marine species that she discovered have been named after her, and numerous awards have come her way, including honorary doctorates. In the spring of 1997, in commemoration of its twenty-fifth anniversary, the United Nations Environment Programme (UNEP) created an exhibit entitled "Eyes on the Environment—Twenty-five Women Leaders in Action." Sylvia Earle was among those cited for her outstanding efforts.

Earle pleads for more, and more strictly regulated, marine reserves in which sea life can recover and for establishing and enforcing international agreements governing human activities in the planet's oceans.

Toward the end of her book, *Sea Change*, she reminds the reader: "Most individuals who change the course of history are not recognized publicly for their good or bad deeds. Most decisions that matter, in fact, seem too trivial to be noticed, but each and every choice made by all the world's individuals, taken together at the end of a day, week, year, or millennium shape the direction of society, cause the actions that collectively build or destroy civilizations."[16]

KATHARINE ORDWAY (1899–1979)
Prairie Guardian

Because she insisted on anonymity, few people knew until after Katharine Ordway's death the extent of her gift to wild fauna and flora—the gift of expansive, unspoiled habitat essential for their survival. The historic contribution to conservation made by this shy and physically frail woman has remained almost a secret—scarcely noted in the annals of conservation philanthropy. Yet, only the contributions of the much wealthier John D. Rockefeller, Jr., exceeded hers.[17]

At one time the prairie of the Great Plains was North America's largest continuous ecosystem—four hundred million acres—and the most characteristic, covering nearly one-fifth of the continent. Now it has become the most fragmented and the rarest. Unlike savannas, which have a scattering of trees, prairies evolve in climates too dry for trees but too wet to be desert. Three kinds of grasses dominate the Great Plains prairies: tallgrass in the rainier eastern part, shortgrass in the drier western areas, and mixed-grass prairie in between. Bluestem and Indian grass, which grow to be more than eight feet high, are the hallmarks of tallgrass prairies. Less than 1 percent of the continent's original tallgrass prairie land remains.[18]

Half the waterfowl of North America derive from prairie potholes and lakes. Being on the Central Flyway, prairie waters also support hosts of migratory species, including shorebirds. And the prairie serves as nesting habitat for many other bird species, including bobolinks, upland sandpipers, lark buntings, and mountain plovers. During the past few decades, the population of grassland species has diminished by between 25 and 65 percent—the worst record for any North American bird group.[19] Originally, tallgrass prairies were home to eighty species of mammals, small and large, from voles to bison.[20] Wildflower blossoms of every color painted the land from April through September. And pharmaceutical plants—the coneflower *(Echinacea)*, feverwort, Western ragweed, and prairie goldenrod, to name a few—flourished.

While growing up in Minnesota, Katharine Ordway loved to watch the prairie's unbroken sea of waving grass and to study the plants and animals inhabiting it. Distress over the vanishing of the prairie compelled her to do something about it, although it was not until the last thirteen years of her eighty-year life that she had the means to act. In her fifties she inherited a substantial fortune from her father, a founder of Minnesota Mining and Manufacturing—now called 3M. She also gained a feeling of purpose and self-confidence that had been lacking earlier while she lived in the shadow of four brothers.

A cum laude graduate of the University of Minnesota with degrees in botany and art, she had also attended Yale Medical School before giving up the idea of a medical career. In her fifties she returned to school to obtain a degree in biology and land-use planning from Columbia University.

Although not an outdoors enthusiast as such—not a hiker or canoeist, nor even much of a birder—Ordway loved nature. Small, thin, and with sensitive skin (she always carried a parasol while in the sun), she left Minnesota because of the harsh winters and established her primary home in the wooded hills of southwestern Connecticut. She spent winters near the Arizona desert and summers at her Long Island home among the dunes and marshes.

In 1959 Ordway established a personal grant-making foundation. Wishing to maintain her privacy, she avoided associating her name with it, calling it the Goodhill Foundation after the road leading to her Connecticut home. With her primary interest being preserving open land and keeping it wild by buying large undeveloped tracts, she sought an entity through which to funnel her foundation's grants and achieve this goal. Her longtime friend and advisor, Richard Pough, a conservation matchmaker, brokered a relationship for her with the Nature Conservancy (TNC), then a fledgling group of which Pough was a founder and officer. A bird-guide author, Pough had also worked for the National Audubon Society for twelve years.

The Nature Conservancy was the only national organization committed exclusively to preserving undisturbed tracts of natural land as examples of ecotypes—distinct animal and plant communities. Of all the types of land least protected and most in need of preservation, TNC ranked prairies at the top of the list. A number of efforts by conservationists in the twentieth century to create national tallgrass prairie preserves had failed because of cattle ranchers' opposition. Doing it privately remained the only option. So TNC compiled a list of suggested prairie properties for Ordway's consideration.

She chose as her first tallgrass prairie preserves several Minnesota sites, including an undisturbed piece in the western part of the state—a relatively small tract that had never been plowed, containing some parts that had never even been mowed or grazed by cattle. She chose another parcel because it had been part of the shrinking range of the greater prairie chicken. Vanishing habitat had driven this grassland bird close to extinction. Since she created this preserve, the greater prairie chicken has returned.

It soon became evident to Ordway that surviving prairies in other states also needed protection, and she underwrote acquisition of small preserves in Nebraska, South Dakota, Kansas, and Missouri. But she was not satisfied with only small preserves. She wanted to protect a vast expanse of tallgrass prairie, one where a person could go and see nothing else—no power lines, houses, or roads—nothing but prairie grass stretching to the horizon.

Though this took years of work by her friends at TNC, in the 1970s such a preserve in the Flint Hills of Kansas became a reality. The Flint Hills run north to south across eastern Kansas, along the western edge of the original tallgrass prairie. Because of their steep and rocky nature, they escaped cultivation. According to TNC this is the only place where thousands of acres of natural tallgrass still exist. Called the Konza Prairie Research Natural Area, Ordway's preserve encompasses over eighty-five hundred acres. Now managed by Kansas State University, it is an outdoor laboratory for long-term ecological research.

Now driven by her sense of purpose, Ordway funded the acquisition of large prairie preserves elsewhere in the Plains states—in Missouri, the Dakotas, and Nebraska. She toured each parcel before approving its purchase. She also requested that all the prairie preserves created with her funding in western Minnesota be given American Indian names; among these are Wahpeton, Pawhuska, Santee, Chippewa, and Osage. At the time of her death in 1979, the Ordway Prairie Preserve System comprised more than thirty-one thousand acres in five midwestern states—the largest sanctuary system of its kind in the world.[21] Since then TNC has

PHOTO: WALLACE C. DAYTON

KATHARINE ORDWAY AND GEOFFREY BARNARD OF THE NATURE CONSERVANCY AT SAMUEL ORDWAY PRESERVE IN SOUTH DAKOTA.

established an additional forty-thousand-acre tallgrass prairie preserve in the southern Flint Hills. Bison, once approaching extinction, now thrive in these rescued preserves along with many other wildlife species.

Sometimes one hears, What good is a prairie if not to be plowed and converted to farmland? The rich soils of the North American prairies, which have been feeding the world, were produced by a complex ecosystem, mostly out of sight underground, storing fertility for thousands of years. Each acre of grasslands soil stores twice as much carbon as forest soil. Roots penetrate downward between twelve and twenty feet,[22] and they are so tightly interwoven that settlers constructed houses with bricks cut from this living carpet. The tough underground stems of prairie tallgrass produced soil so dense it required turning with a special sod-breaking plow before it could be tilled. The invention of this plow led to the destruction of most of the tallgrass prairies.

Scientists from far-flung institutions now participate in the National Science Foundation's Long-Term Ecological Research Program on the Konza preserve. They seek to find out what makes the tallgrass prairie such a long-lasting system by examining, among other things, the complex interactions among the invertebrates and other organisms in the plants' root systems. The goal is to develop ways to enhance the management of wildlife, water, and grassland.

Ordway did not stop with preserving prairie lands. She and her foundation also funded the acquisition of wilderness areas, barrier islands, tidal marshes, scrub forests, an unaltered palmetto prairie in Florida, and a significant Pacific Flyway link in Oregon, which is also an important nesting area for threatened bird species. After her death, her Goodhill Foundation contributed $11 million to save forest and prairie habitat along twenty-two miles of the Niobrara River in northern Nebraska—the last long stretch of wild river in the Great Plains.

Ordway had not used much of the money in her foundation for contributions while she lived, choosing instead the thrill of giving directly from her personal resources. She instructed her board of directors to distribute the foundation's assets ($53 million) quickly after her death to acquire natural areas for preservation. They did so over the next five years, multiplying the money's effectiveness through challenge grants. Ten million dollars went toward a Nature Conservancy program to save remaining representative ecosystem types—from New England to Hawaii—that were at risk of being lost forever. This program (in two phases) resulted in the acquisition of 192,000 acres of rare natural areas threatened with development.

The Goodhill Foundation also funded TNC's Katharine Ordway Endangered Species Conservation Program, assisting projects in thirty-two states to safeguard habitat for over one hundred of the most endangered species of flora and fauna, including bat caves in Kentucky.

Ordway, an art collector and world traveler, never married. She lived a quiet and undramatic life. But she left a formidable legacy to wildlife. Directly and through her Goodwill Foundation, she contributed at least $64 million to conservation. Her example inspired other Ordway family members to support conservation programs. The $53 million channeled to TNC over twenty years also helped the organization evolve into an effective leader in saving natural diversity.

The United Nations Educational, Scientific and Cultural Organization designated Ordway's Konza Prairie and her Atlantic barrier-island sanctuaries as International Biosphere Reserves—part of a worldwide network of ecosystem examples focused on scientific research and conservation.

[B]ecause women are the repository of the largest store of unused abilities they will surely be a primary source and resource of ideas and energy.

ELIZABETH JANEWAY
Between Myth and Morning

Women and International Forums

BELLA ABZUG (1920–1998)
Creates a Global Linchpin

If Bella Abzug were to have had her way, women will run the twenty-first century, bringing vital changes and correcting the "errors of the past, which have been created largely by one part of the population. . . . You can't continue to have a world without equal participation of men and women. That's my central thesis. Sometimes I say that it's not that I think women are superior to men, it's just that we've had so little opportunity to be corrupted by power. And I jokingly add sometimes that we want that opportunity."[1]

During the 1990s, in her continuing endeavor to empower women, Abzug built a worldwide movement, the Women's Environment and Development Organization (WEDO). It provides a support structure for women's nongovernmental organizations from both Northern and Southern countries. She shared the leader's role with ten other women from Brazil, Kenya, Costa Rica, India, Guyana, Norway, Egypt, Nigeria, New Zealand, and the United States. The group's stated vision is to enhance women's leadership and advocacy skills in order to help them transform their concerns about the environment, development, population, and gender equity into programs and policies in countries around the world.

WEDO was born in 1990 when Abzug and Mim Kelber—a friend from Hunter College days—created the coalition to monitor United Nations activities and to give women more visibility. In 1991 Abzug learned that the proposed action documents to be considered at the UN's 1992 Earth Summit in Brazil failed to address the critical connections between women's roles and the planet's environmental crisis. Shocked by this information, she and colleagues from throughout WEDO's network convened an independent women's conference in November 1991 to develop an alternative action proposal reflecting women's perspectives.

This Women's Congress for a Healthy Planet—over which Abzug presided—drew fifteen hundred women from eighty-three countries to Miami. As mentioned earlier in this book, the document they hammered out—called the Women's Action Agenda 21, and reflecting a consensus of North and South—influenced not only the Earth Summit's action plan but subsequent UN conference agendas as well. Among these were the conferences on population (in Cairo), where Abzug served as an official delegate, human rights (in Vienna), and women (in Beijing). Concurrently with each UN conference, WEDO coordinated a parallel women's forum and convened daily caucuses to track and influence deliberations of the official delegates. During the preparations for the Earth Summit in 1992, Abzug, as head of WEDO, served as senior advisor to Maurice Strong, secretary-general of the summit. WEDO also established its own fifty-one-member advisory council of environment and development experts.

Using the experience and methods derived from their parallel forum at the Earth Summit, WEDO assembled a resource manual for nongovernmental organizations on how to bring women's perspectives into UN deliberations. The publication is called *Women Making a Difference*. In her preface, Abzug observed:

> In my lifetime, I have witnessed remarkable changes and the impressive power of individual and group action in concert with others. The story of the fast-growing global women's movement is among the most inspiring.

It is a work in progress, drawing its vision and sustenance from women working in their own communities and nations as well as at the international level. Mothering Earth will take many hands and minds, with women and men working together as equal partners to assure the people of the world and future generations a healthy, liveable and socially just planet.[2]

And in an article for a UN publication, Abzug remarked: "Everywhere, women are catalysts and initiators of environmental and democratic activism, demanding an equal voice in 'fate of the Earth' decisions and an end to the poverty and exploitative development patterns that often primarily victimize women and children. . . . Women have moved into the forefront of environmental initiatives because they are typically the first to recognize environmental problems and are most directly affected by them."[3] She added that women want to be represented in parliaments and congresses, sharing power in equal numbers with men. If they can be present in this way at all decision-making tables, women will provide an infusion of new elements and an alternative perspective, which Abzug believed is essential to achieving sustainable development.

Her half century of citizen activism before 1990 in causes such as civil rights, disarmament and world peace, equal rights for women, defending victims of Senator Joseph McCarthy, and opposing the Vietnam War sharpened Bella Abzug's organizing, speaking, and leadership skills. She also wrote three books and was a news commentator for a time. After being appointed by President Jimmy Carter as presiding officer of the Women in Power Committee of the National Commission on the Observance of International Women's Year in 1975, she was fired in 1979 because of her view that women worldwide should ignore national boundaries and unite. Through WEDO in the 1990s, she made that vision a reality.

Abzug grew up in the Bronx, a borough of New York City. After graduating from the Columbia University School of Law in 1947 (she was one of only six women students), Abzug, by then married, practiced law in New York City. She often represented "the underdog" or indigent African Americans in the South on a pro bono basis. Once—and when she was pregnant—she had to sleep in a southern bus station because no local hotel would give her a room. Since women lawyers were a rarity in the 1940s and 1950s, judges often mistook her for a legal secretary. Abzug resorted to wearing broad-brimmed hats to mark her distinction—an accessory that became her fashion trademark.

In 1972 Abzug ran for Congress on a women's-rights and peace platform and became the first woman to win a seat on such a basis. During her six years in the U.S. House of Representatives, she helped found the National Women's Political Caucus, drafted the landmark Freedom of Information Act, and supported the development of alternative, renewable energy sources in opposition to nuclear power. In 1975 she served as congressional advisor to the U.S. delegation to the UN Conference on the Decade for Women in Mexico City.

Described as blunt, aggressive, and brash, she never hesitated to reprimand presidents or defy precedents. Breaking with House tradition, she stood up on her first day in Congress to propose a resolution to withdraw troops from Indochina. And she was the first representative to demand the impeachment of President Richard Nixon. Abzug did not mind being called aggressive; in a world dominated by men, she said, boldness is the only way to gain a voice.

In 1985, extending her reach, Abzug formed the Women's Foreign Policy Council, whose objective was to make women more visible in this area of decision making. The council put together a list of women considered to be experts on various aspects of foreign affairs. Efforts like these, and the involvement of the Women's Environment and Development Organization in creating international forums during the 1990s, helped smooth the way for women in foreign policy. President Clinton's appointment in 1997 of the first woman to be secretary of state marks a milestone for women's increasing visibility and participation in foreign affairs.

To Bella Abzug, a healthy global environment encompassed more than high-quality air, water, wildlife habitat, forests, and soil. She said it also included economic rights, human rights, peace, justice, human health, and shared power in decision making. Male-dominated decision making has resulted in what she called a global nervous breakdown. The problems demand, she believed, that the collective wisdom, knowledge, and experience of women be brought to bear on the decision-making process.

One of WEDO's ongoing initiatives involves monitoring and prodding various nations regarding their fulfillment (or lack thereof) of commitments made at the UN's international conferences of the 1990s. Abzug set in motion a campaign called Contract with Women of the USA as a follow-up to the UN's Fourth World Conference on Women, held in Beijing. Even though the platform adopted at that event is not *legally* binding, women of the world consider it politically binding. They are lobbying to translate the promises into policies.

In the meantime, WEDO supports training programs on site for local community activists in the Americas, Africa, Asia, Europe, and the Caribbean. The group has also brought teams of women from developing countries to New York and Washington, D.C., for training as activists and to strengthen their skills as local leaders. And in 1997, under Abzug's direction, WEDO launched a major initiative to link environmental contaminants to human health problems—for example, breast cancer.

At the September 1995 UN Conference on Women in Beijing, the United Nations Development Programme honored Abzug for her pioneering work in the women's movement. In her acceptance remarks, Abzug pointed out:

> We must elect more women—yes. But we must transform those structures to which we elect women to accomplish our goals, because present institutions will not do. . . . Our struggle is about reversing the trends of the social, economic and ecological crisis. . . . Our struggle is about creating sustainable lives and attainable dreams. . . . For us to realize our dreams, we must keep our heads in the clouds and our feet on the ground. . . . In my heart I believe that women will change the nature of power rather than power change the nature of women.[4]

Can we identify any root cause of the world's environmental predicament? Could it have resulted from our not having an adequate ethical and philosophical foundation for our activities?

If there is a single ecological lesson to be learned from the history of the past two centuries, it may be that we desperately need to establish such underpinnings. At the least, it would be wise to adopt the "precautionary principle" in all decisions involving science and technology.[1] In other words, any activity that might involve an unacceptable level of risk, that might result in a hazardous outcome, or whose long-term effects are not known should not go forward.

Preventing environmental degradation rather than having to restore damaged ecosystems makes good sense. The absence of scientific certainty about environmental damage cannot be an excuse for failing to use a worst-case scenario in decision making. Although this approach is ignored by most countries, and holds little hope of short-term economic benefits, the survival of our species and of all life forms depends on managing Earth's complex and interdependent life-supporting ecosystems in sustainable ways. We ignore the precautionary principle at our peril.

Saving the planet, in other words, calls for a profound transformation in the heart, mind, and spirit of human beings. Fortunately, this transformation is already under way, giving rise to an emerging "Integral Culture." Forty-four million leading-edge thinkers (or 24 percent of U.S. adults), called "Cultural Creatives" by the sociologist Paul H. Ray, spearhead this movement, a group comprising 50 percent more women than men.[2] Such trends give us hope that change can happen before it is too late. Perhaps it is no surprise that women are leading the way.

PREFACE

1. Bruce Hozeski, *Hildegard of Bingen: The Book of the Rewards of Life* (New York: Garland Publishers, 1993), 137–39.
2. Ashley Montagu, *The Natural Superiority of Women* (New York: Macmillan, 1953), 180.
3. Sy Montgomery, *Walking with the Great Apes* (Boston: Houghton Mifflin, 1991).

CHAPTER ONE

1. Thomas Weber, *Hugging the Trees: The Story of the Chipko Movement* (New York: Penguin Press, 1989), 91.
2. One-third of the earth's tillable land has already been lost to erosion during the past four decades. Once eroded, soil takes thousands of years to restore itself. See Bhaskar Nath, Luc Hens, and Dimitri Devuyst, eds., *Sustainable Development* (Brussels: VUB University Press, 1996), 92.
3. "Devi" and "Behn" are not surnames, but terms commonly attached to given names in India to show respect (Devi) or sisterhood (Behn). Hence, they do not connote blood relationships.
4. Vandana Shiva, *Staying Alive—Women, Ecology, and Development* (London: Zed Books, 1988). Shiva herself grew up in the Himalayan forests. She became a Chipko volunteer activist involved in, among other things, writing articles to support the movement. (See chapter 9 of this book.)
5. Brian Nelson, "Chipko Revisited," *Whole Earth Review*, summer 1993, 116.
6. Until the British colonial administration usurped it, the Himalayan land had been managed as common property by its inhabitants.
7. Sarala Behn, "A Blueprint for Survival of the Hills," supplement to *Himalaya: Man and Nature* (New Delhi: Himalaya Seva Sangh, 1980).
8. The Right Livelihood Award was established by Jakob von Uexkull to honor and support those working on exemplary and practical solutions to contemporary

world problems, thereby contributing to making life more whole and healing the planet.

9. Right Livelihood Foundation, press release, Stockholm, 9 October 1987.

10. Wangari Maathai, "Kenya's Green Belt Movement: A Community-Based Project Created and Directed by Women," *UNESCO Courier*, March 1992, 23.

11. Ibid., 25.

12. Quoted in Jeremy Seabrook, *Pioneers of Change: Experiments in Creating a Humane Society* (Philadelphia: New Society Publishers, 1993), 51.

13. Wangari Maathai, "Foresters without Diplomas," *Ms.*, March–April 1991, 74.

14. Aubrey Wallace, *Eco-Heroes: Twelve Tales of Environmental Victory* (San Francisco: Mercury House, 1993), 135.

15. Forests once covered more than 40 percent of the earth's land surface; the figure today is only 27 percent. Anjali Acharya, "Forest Loss Continues," in *Vital Signs, 1996: The Environmental Trends That Are Shaping Our Future*, ed. Lester Brown (Washington, D.C.: Worldwatch Institute; New York: W. W. Norton, 1996), 122.

16. The spotted-owl brouhaha in Oregon generated predictions by the timber industry that setting aside forest habitat for an endangered species would destroy the state's economy through loss of jobs. The opposite happened. Oregon saw its lowest unemployment rate in a generation. The number of new jobs arising in high-tech industries, professional services, and tourism exceeded those lost in timber enterprises. Bill McKibben, "What Good Is a Forest," *Audubon*, May–June 1996, 56.

17. Rainforest Action Network, "Clayoquot Rainforest Coalition Overview," Internet posting, January 1997.

18. Wallace, *Eco-Heroes*, 148.

19. Ibid., 151.

20. "A forest's watershed protection value alone can exceed the value of its timber." Janet Abramovitz, "Valuing Nature's Services," in *State of the World, 1997*, ed. Lester Brown (Washington, D.C.: Worldwatch Institute; New York: W. W. Norton, 1997), 96.

21. Rainforest Action Network, "International Rainforest Summit Condemns British Columbia's Democracy-Slam in Slocan Valley," Internet press release, London, 5 September 1996.

22. In 1996, California environmentalists launched a shareholder campaign to pressure Pacific Bell to convert to recycled and nontree fibers for its phone books. Seven California communities have passed resolutions condemning the company's use of paper made from British Columbia's prime rain forest trees. As of March 1996, more than thirty-five thousand Pacific Bell customers had communicated with the company, asking it to stop the practice. Rainforest Action Network, "Yellow Pages Controversy Goes to Pacific Bell Stockholders," Internet press release, 21 March 1996.

23. The United States ranks fifteenth worldwide in paper recycling. Alice Horrigan and Jim Motavalli, "Talking Trash," *E, the Environmental Magazine*, March–April 1997, 28. Another writer notes that "[t]he United States consumes twice as much wood as other industrial countries, the equivalent of one mature tree per person per year." Vince Bielski, "Shopper, Spare That Tree," *Sierra*, July–August 1996, 39.

24. Harry R. Carter, "The Marbled Murrelet: How Did These Little-Known Seabirds Become a Symbol for Saving Old-Growth Forests?" *Tideline*, newsletter of San Francisco Bay Wildlife Refuge, Winter 1995, 1–3. See also Jane Kay, "Fighting for the Tall Trees: Part One Sidebar—The Case of the Marbled Murrelet," *San Francisco Examiner*, 16 December 1995. (The Maxxam/Pacific Lumber Company intimidated its own biologists into covering up their data.)

25. Earth First!'s statement of purpose says: "Non-violent direct action at the point of production (or in this case, the point of destruction) is the main strategy of Earth First! in the Headwaters campaign. In the tradition of . . . Martin Luther King Jr. and Ghandi [*sic*], we believe in resistance to injustice and ecological destruction through civil disobedience by putting our bodies on the line." Earth First! Web site on the Internet.

26. Spiking and monkey-wrenching continued among some Earth First! groups elsewhere in the United States, driving away many activists.

27. Darryl Cherney, Judi Bari, and Northcoast California Earth First!ers, "Tree-Spiking Renunciation and Mississippi in the California Redwoods," Internet memorandum, April 1990, Earth First! Web site.

28. A ten-year veteran of the mills who had joined with Bari to get his company cited for "willful exposure of employees" was fired.

29. By odd coincidence, the rifle scope and crosshairs is the assassination symbol of the Secret Army Organization, according to Bari an FBI-funded right-wing paramilitary group in San Diego organized in the 1970s by Richard Held, the FBI agent put in charge of investigating the bombing of Bari's car.

30. Judi Bari, "Misery Loves Company," *Anderson Valley (California) Advertiser*, 22 May 1991.

31. Judi Bari, "Community under Siege," speech at the Cinco de Mayo celebration, Booneville, Calif., 5 May 1991. Reprinted in Judi Bari, *Timber Wars* (Monroe, Maine: Common Courage Press, 1994), 128.

32. Richard Held resigned his job as head of the FBI's San Francisco office on the same day that the photos of the bombed car were released.

33. Judi Bari interview, *New Settler*, no. 89 (Mendocino, Calif., 1995).

34. Bari, *Timber Wars*, 123.

35. Judi Bari, "The Bombing Story—Part 1. The Set-Up," *Earth First! Journal*, 1 May 1994. Reprinted in Bari, *Timber Wars*, 286.

36. Bari, *Timber Wars*, 123.

37. Ibid., 130.

38. Judi Bari, "Headwaters Forest Still Stands! Mass Protest Averts Logging Threat (For Now)," Mendocino Environmental Center, Ukiah, Calif., press release, mid-September 1995.

39. Ibid.

40. Judi Bari, "An Open Letter to *ABC Network News* from the *Earth First! Journal*," 7 April 1996 (Internet).

41. Daniel Sneider, "Species Act Survives Challenge—For Now," *Christian Science Monitor*, 21 February 1997.

42. In 1997, a Justice Department investigation of the FBI's crime lab exposed it as

having mishandled evidence, bungled cases, and misled the courts. Daniel Klaidman and Peter Annin, "Under the Microscope: The Once Legendary FBI Crime Lab Is Swamped by Charges of Sloppiness," *Newsweek*, 10 February 1997, 32.

43. Judi Bari, "Judi Bari vs. FBI Car-Bomb Lawsuit Hearing, Part 1: Evidence to be Revealed," Redwood Summer Justice Project, Santa Rosa, Calif., November 1996 (Internet).

44. In the late 1990s, concern about the continuing decline of salmon in Washington and Oregon led to serious consideration of removing four major dams on the Snake River, a major tributary of the Columbia. According to a report commissioned by the U.S. Army Corps of Engineers, this would be the surest, but the most expensive, way to save the endangered chinook. Brad Knickerbocker, "Saving an Icon of the Pacific Northwest," *Christian Science Monitor*, 30 January 1997.

45. "Research done by the U.S. Forest Service indicates that recreation yields four times the economic benefits of timber and creates sixteen jobs for each one derived from cutting trees." David Seideman, "Out of the Woods," *Audubon*, July–August 1996, 70.

46. Marilyn Fike, administrator, Bullitt Foundation, Seattle, by e-mail to the author, 4 April 1997.

47. For several years in the late 1980s and early 1990s, Janet Brown was the unpaid chairwoman of Friends of the Earth, an NGO with a two-million-dollar budget and thirty thousand members.

48. World Wildlife Fund, "First-of-Its-Kind World Forest Map Reveals 94 Percent of Earth's Forest Have No Formal Protection," Internet press release, Washington, D.C., 9 September 1996.

49. The Harvard entomologist Edward O. Wilson says of insects: "The truth is that we need invertebrates but they don't need us. If human beings were to disappear tomorrow, the world would go on with little change. . . . [B]ut if invertebrates were to disappear, it is unlikely that the human species could last more than a few months." Edward O. Wilson, *In Search of Nature* (Washington, D.C.: Island Press, 1996), 144.

50. Donella Meadows, *Global Citizen* (Washington, D.C.: Island Press, 1991), 79.

51. Al Gore, *Earth in the Balance: Ecology and the Human Spirit* (New York: Penguin Books, 1993), 116, 119.

52. Conservation International, Internet Web site, February 1997.

53. Kathryn Fuller, luncheon speech before WorldWIDE's Washington, D.C., Forum, 12 March 1991, *WorldWIDE News*, March–April 1991.

54. Kathryn Fuller, Foreword to *World Wildlife Fund Annual Report, 1996* (Washington, D.C.: World Wildlife Fund, 1996), 8.

CHAPTER TWO

1. William Ripley Nichols, "The Present Condition of Certain Rivers of Massachusetts Together with Considerations Touching the Water Supply of Towns," *Fifth Annual Report* (Boston: Massachusetts State Board of Health, January 1874), 64.

2. Ellen Swallow Richards, *Conservation by Sanitation* (New York: John Wiley and Sons, 1911), 107.
3. Ellen Swallow Richards, *Air, Water, and Food from a Sanitary Standpoint* (New York: John Wiley and Sons, 1901), 43–44.
4. As recently as 1989 food was still being hauled around the country in contaminated trucks and in rail cars infested with maggots. The Associated Press reported that "[t]rucks and rail cars are hauling garbage into America's breadbasket and then carrying food on the way back with little or no cleaning in between, according to testimony on Capitol Hill." Robert Greene, "House Panel Hears of Trucks That Haul Garbage, Then Food," *Wilmington (Delaware) News Journal,* 10 November 1989.
5. U.S. Department of Commerce, Bureau of the Census, *Historical Statistics of the U.S., Colonial Times to 1970,* Series B 201–213 (Washington, D.C., 1975), 63.
6. Since the passage of major environmental laws by the U.S. Congress in 1970, some progress has been made in cleaning up our environment. However, even as we enter the last few years of the century, we still have a long way to go.

   "Millions of pounds of toxic chemicals, like lead, mercury and pesticides, pour into our waterways each year contaminating wildlife, seafood and drinking water; one-half of our nation's lakes and one-third of our rivers are too polluted to be completely safe for swimming and fishing; raw sewage, poison runoff and other pollution have caused 8,000 beach closures or advisories over the past five years; Americans are exposed to 70,000 chemicals, some 90 percent of which have never been subjected to adequate testing to determine their impact on our health; as of 1994, 1.7 million American children, ages one to five, suffered from lead poisoning." Natural Resources Defense Council, *Annual Report, 1995* (New York: NRDC, 1995), 10, 14.
7. Richards, *Conservation by Sanitation,* 86, 87, 107.
8. Ellen Swallow Richards, *Annual Report,* Society to Encourage Studies at Home, c. 1880.
9. Ellen H. Richards, *Sanitation in Daily Life* (Boston: Whitcomb and Barrows, 1907), 33.
10. Ibid., v.
11. Ibid., 25.
12. Speech by Ellen Swallow Richards before the American Public Health Association Annual Meeting, Boston, 1896. American Public Health Association archives, Washington, D.C.
13. But more than one hundred years later the NRDC observed that "[w]hile protected by a network of health and safety laws, school buildings are not necessarily safeguarded for children. In fact, children—who are more vulnerable to the effects of pollution—may spend up to forty hours a week in buildings that are more densely occupied than many offices and that house health hazards such as collapsing walls, pesticide spills, lead and asbestos contamination, and polluted indoor air." "ABCs of Safe Schools," *Amicus Journal,* winter 1996, 7.
14. Ellen H. Richards, *Food Materials and Their Adulterations* (Boston: Estes and Lauriat, 1886).

15. The first Federal Food and Drug Act did not become law until 1906; *Guide to American Law, Everybody's Legal Encyclopedia* (New York: West Publishing, 1984), 4:356.

16. "New Science: Mrs. Richards Names It Oekology," *Boston Daily Globe*, 1 December 1892, 1.

17. "The Inauguration of Marion LeRoy Burton, Ph.D., D.D., as President of Smith College," 5 October 1910 (Northampton, Mass.: Smith College, 1911), 67.

CHAPTER THREE

1. I rely throughout this chapter on Martin Melosi, ed., *Pollution and Reform in American Cities, 1870–1930* (Austin: University of Texas Press, 1980); on Suellen M. Hoy, "Municipal Housekeeping: The Role of Women in Improving Urban Sanitation Practices, 1880–1917," in Melosi, *Pollution and Reform;* and on Robert McHenry, ed., *Liberty's Women* (Springfield, Mass.: G and C Merriam, 1980).

2. Al Gore, *Earth in the Balance: Ecology and the Human Spirit* (New York: Penguin Books, 1993), 155.

CHAPTER FOUR

1. Richard L. Kenyon, "Assault on Nature," editorial, *Chemical and Engineering News*, 23 July 1962, 5.

2. In 1962—the same year that Rachel Carson's history-changing book, *Silent Spring*, hit the best-seller list—the thalidomide tragedy stunned the world. This sleep-inducing drug, which had passed safety tests on animals with flying colors, left twelve thousand newborns in forty-six countries severely deformed. Americans escaped only because Frances Kelsey, a stubborn and courageous woman physician on the staff of the Food and Drug Administration, had refused—in the face of intense pressure from the pharmaceutical industry—to approve thalidomide until additional safety data became available.

3. Biology, "Pesticides: The Price of Progress," *Time*, 28 September 1962, 45.

4. In a speech before the Women's National Press Club, Washington, D.C., 5 December 1962, Carson remarked: "My text this afternoon is taken from the *Globe Times* of Bethlehem, Pa., a news item in the issue of October 12. After describing in detail the adverse reactions to *Silent Spring* of the farm bureaus in two Pennsylvania counties, the reporter continued, 'No one in either county farm office who was talked to today had read the book, but all disapproved of it heartily.' " Manuscript, Beinecke Rare Books and Manuscripts Library, Yale University, New Haven, Conn.

5. Quoted in Frank Graham, *Since "Silent Spring"* (Boston: Houghton Mifflin, 1972), 23.

6. David A. Birk, "IPM ABC's," *Audubon Journal* (Delaware Audubon Society), January–February 1996, 1.

7. Talk of the Town, *New Yorker*, 26 May 1945, 18.

8. Theo Colborn, Dianne Dumanoski, and John Peterson Myers, *Our Stolen Future: Are We Threatening Our Fertility, Intelligence, and Survival?—A Scientific Detective Story* (New York: Dutton, 1996), 89.

9. Quoted in Paul Brooks, *The House of Life: Rachel Carson at Work* (Boston: Houghton Mifflin, 1972), 244.

10. Rachel Carson, *Silent Spring* (Boston and New York: Houghton Mifflin, 1962), 8, 13.

11. Paul R. and Anne H. Ehrlich, *Healing the Planet* (New York: Addison-Wesley Publishing, 1995), 204.

12. Graham, *Since "Silent Spring,"* 49–50.

13. Brooks, *The House of Life*, 18n.

14. Carson, Women's National Press Club speech.

15. Rachel Carson, speech before the Women's National Book Association, New York, 15 February 1963. Manuscript, Beinecke Rare Books and Manuscripts Library, Yale University, New Haven, Conn.

16. Thomas N. Schroth, ed., *Congress and the Nation—1945–1964: A Review of Government and Politics* (Washington, D.C.: Congressional Quarterly Service, 1965), 1183.

17. In the mid-1980s women in Sweden offered jam made from tainted berries to members of Parliament in protest against the use of herbicides on forests. Carolyn Merchant, *The Death of Nature: Women, Ecology, and the Scientific Revolution* (San Francisco: Harper, 1989), xv.

18. Jennifer D. Mitchell, "Nowhere to Hide: The Global Spread of High Risk Synthetic Chemicals," *World-Watch*, March–April 1997, 26.

19. Gary Gardner, "Organic Farming Up Sharply," in *Vital Signs, 1996: The Environmental Trends That Are Shaping Our Future*, ed. Lester Brown (Washington, D.C.: Worldwatch Institute, and New York: W. W. Norton, 1996), 110.

20. Toni Nelson, "Efforts to Control Pesticides Expand," in Brown, *Vital Signs, 1996*, 109.

21. Archives, Rachel Carson Council, Washington, D.C., August 1980.

22. Other members of the panel were Russell Baker, Rosamond Bernier, Daniel J. Boorstin, John Cage, Judy Chicago, William Sloane Coffin, Robert Coles, Ellen V. Futter, John Kenneth Galbraith, Nikki Giovanni, Theodore M. Hesburgh, Maxine Hong Kingston, Margaret E. Kuhn, Frances Lear, Eugene J. McCarthy, Sandra Day O'Connor, Linus Pauling, Wendy Wasserstein, George F. Will, Tom Wolfe, and Eugenia Zukerman.

23. Opinion polls taken between 1965 and 1970 reflect *Silent Spring*'s influence on public attitudes. In 1965 only 28 percent of those questioned considered air pollution "very serious." By 1970, the percentage had risen to 69 percent. Over the same five years, concern about the seriousness of water pollution rose from 35 percent to 74 percent. Hazel Erskine, "The Polls: Pollution and Its Cost," *Public Opinion Quarterly* (spring 1972): 121.

24. Graham, *Since "Silent Spring,"* 74.

25. The endocrine system comprises a set of hormone-producing organs—the brain,

pituitary and adrenal glands, sex organs, liver, and pancreas. Through a complex series of feedback mechanisms involving the immune system, the central nervous system, and these organs, the endocrine system controls hormones released into the bloodstream. These function as chemical signals to any place in the body that needs a particular cellular action. The precision and timing of the hormonal signals are particularly delicate during fetal development.

26. Theo Colborn, "Toxic Chemicals—We Are All at Risk," speech before the State of the World Forum, San Francisco, 3 October 1996. (State of the World Forum Web site on the Internet.)

27. "[A] study by the Foundation for Advancements in Science and Education examining U.S. pesticide exports between 1992 and 1994 found that at least 7,000 tons (15 million pounds) of banned pesticides were exported, with some heading to nations where their use is also forbidden." Nelson, "Efforts to Control Pesticides Expand," 109. The exports of banned and unbanned pesticides during that period amounted to 114,600 tons. Ted Williams, "Silent Scourge," *Audubon*, January–February 1997, 28.

28. Theo Colborn, "Scientists Say the Darndest Things," speech at a forum sponsored by the International Platform Association, Washington, D.C., 3 August 1996 (Internet posting).

29. Ibid.

30. Colborn, Dumanoski, and Myers, *Our Stolen Future*, 13.

31. Ibid., 111.

32. Colborn, "Scientists Say the Darndest Things."

33. "Leaked memos and other internal documents made available to Sierra show that the industry has known for some time about the reproductive and other health risks associated with chlorinated chemicals, and has been deeply worried about the public awareness of the issue that the new book could generate." David Helvarg, "Poison Pens," *Sierra*, January–February 1997, 31. One public relations firm suggested the industry use women as spokespersons to help counteract the book's revelations. Ibid.

34. Colborn, "Scientists Say the Darndest Things."

35. "Statement from the Work Session on Environmental Endocrine Disrupting Chemicals: Neural, Endocrine, and Behavioral Effects," report on work session held in Erice, Sicily, 5–10 November 1995; see also *Christian Science Monitor*, 31 May 1996.

36. Colborn, "Toxic Chemicals."

37. Ibid.

38. Colborn, "Scientists Say the Darndest Things."

39. "Taking Back Our Stolen Future: Hormone Disruption and PVC Plastic," Greenpeace Web site, Internet posting, April 1996.

40. Donella Meadows, *Global Citizen* (Washington, D.C.: Island Press, 1991), 136.

41. The Price-Anderson Indemnity Act was passed in 1957 to shield the nuclear industry from the enormous costs that would result from a nuclear accident. It was demanded by the industry as a condition for private utilities' building nuclear plants. It provides, in place of insurance policies taken out by individuals, for

$500 million in federal coverage for nuclear accidents. The utilities were to provide $60 million through joint insurance pools. The act has been renewed in every decade since.

42. The Palisades plant is in southwestern Michigan, about a hundred miles north of Chicago.

43. Point Beach is on the western shore of Lake Michigan, near Appleton, Wisconsin.

44. Patricia Adams and Philip Williams, introduction to *Yangtze! Yangtze!: Debate over the Three Gorges Project*, ed. Dai Qing, ed. Patricia Adams and John Thibodeau, trans. Nancy Liu et al. (Toronto and London: Probe International and Earthscan, 1994), xxiii.

45. Audrey Ronning Topping, "Dai Qing, Voice of the Yangtze River Gorges," "Three-River Gorges Controversy," and "What Now for Three-River Gorges?" *Earth Times News Service*, 1996, Internet.

46. Janet N. Abramovitz, "Imperiled Waters, Impoverished Future: The Decline of Freshwater Ecosystems," *Worldwatch Paper 128* (Washington, D.C.: Worldwatch Institute, 1996), 13.

47. International Rivers Network, e-mail, 4 September 1997.

48. Zhu Jianhong, "The Three Gorges Project: An Enormous Environmental Disaster. An Interview with Hou Xueyu," in Dai, *Yangtze! Yangtze!* 189.

49. Abramovitz, "Imperiled Waters, Impoverished Future," 58.

50. Dai Qing, Freeman Lecture, M.I.T., Cambridge, Mass., 16 April 1996, audiotaped and transcribed by the author.

51. John Tuxill, "Past Dam Disaster Casts Shadow over Three Gorges," *World Watch* July–August 1996, 6.

52. In March 1997, one hundred representatives from twenty countries (not including China), at the First International Meeting of Dam-Affected People, demanded an immediate international moratorium on the building of large dams until a number of demands are met. International Rivers Network, Internet press release.

53. Among the results of the Aswan Dam: "The spring floods are gone, the nutrient-rich silts no longer come; the Nile sardine fishery in the Mediterranean is going extinct; bilharzia, or schistosomiasis, a gruesome disease borne by a snail that thrives in slack waters in Africa, is rampant; the reservoir is silting up quite rapidly due to the erosion from primitive agriculture upriver; irrigation canals, meanwhile, are being scoured by the silt-free water released by the dam; and the salts have arrived. Egyptian farmers have been irrigating madly, and the water table, increasingly poisoned by salts, is rising dangerously." Marc Reisner, *Cadillac Desert: The American West and Its Disappearing Water* (New York: Viking Penguin, 1986), 487.

54. Before 1900, only 8,750 kilometers of the world's once free-flowing rivers had been artificially altered for navigation; by 1980, the figure was 498,000 kilometers. Abramovitz, "Imperiled Waters, Impoverished Future," 13.

55. Reisner, *Cadillac Desert, 175, 503.*

CHAPTER FIVE

1. Russell Means, with Marvin J. Wolf, *Where White Men Fear to Tread: The Autobiography of Russell Means* (New York: St. Martin's Press, 1995), 293.
2. Ibid., 375.
3. Tom Wilkinson, "Indian Reservations Reap Deadly Legacy of WW II," *Christian Science Monitor*, 9 December 1996.
4. Means, *Where White Men*, 375.
5. Joan Martin-Brown, "Women in the Ecological Mainstream," in *Women and the Environment: An Analytical Review of Success Stories*, ed. Waafas Ofosu-Amaah and Wendy Philleo (Washington, D.C.: United Nations Environment Programme and WorldWIDE Network, 1992), Appendix 3, 138.
6. Ibid., 138–39.
7. A timber company in California, as part of an attack by corporations on environmental education in schools, tried to have *The Lorax* banned in 1996. Michael Satchell, "Dangerous Waters? Why Environmental Education Is under Attack in the Nation's Schools," *U.S. News and World Report*, 10 June 1996, 63.
8. Martin-Brown, "Women in the Ecological Mainstream," 143–44, 146.

CHAPTER SIX

1. "An Alarming Silence on Chemical Wastes," *Business Week*, 7 May 1979, 46.
2. Lois Gibbs, as told to Murray Levine, *Love Canal—My Story* (Albany: State University of New York Press, 1982), 28. (Further references to this work, abbreviated *LC*, will be included parenthetically in the text of chapter 6.)
3. The New York Department of Environmental Conservation later demolished 239 houses during the containment process. The dump, because of its mixture of so many toxics, can never really be cleaned up. It is, instead, "contained" by a forty-acre landfill and capping system requiring constant surveillance and maintenance. By 1990 most of the work had been completed. It took over a decade.
4. In the mid-1990s, the revitalization agency continues to peddle the 234 renovated houses near the landfill to low-income families at 10 to 15 percent below market value. Buyers must agree to keep houses owner-occupied; the sales-agreement documents disclaim any government warranty of habitability. Andrew J. Hoffman, "An Uneasy Rebirth at Love Canal," *Environment*, March 1995, 5.
5. In 1983 the Centers for Disease Control disputed the EPA's chromosome findings. See "Love Canal Revisited: Still More Questions," *Newsweek*, 30 May 1983, 41.
6. In October 1983, Hooker Chemical's parent company, Occidental Petroleum Corporation, reached an out-of-court settlement with the 1,345 Love Canal residents. It cost the company itself between $5 and $6 million, with insurance paying the remainder of the $25 million settlement. Occidental also reached an out-of-court settlement with the state for $98 million to reimburse New York for

cleanup expenses. Cases involving the federal government were still pending in the mid-1990s. Hoffman, "An Uneasy Rebirth."

7. By 1988 the EPA had entered more than a thousand hazardous-waste sites on its priority cleanup list. The worst polluter is the U.S. government; its facilities discharge two and a half million tons of unreported radioactive and toxic waste. The General Accounting Office reports that 95 percent, or two hundred million tons, of all chemical pollution goes unreported because the government is exempt from reporting, the EPA is weak on enforcement, and the Superfund law is full of loopholes. "Taking on the Nation's Biggest Polluters," *20/20 Vision*, California Sixteenth District, Monterey County, September 1991.

8. *WorldWIDE News*, Winter 1993, 7. Akiko Domoto herself founded Genki, an organization of more than eight hundred Japanese women focusing on environmental issues.

9. Michiko Ishimuri, *Paradise in the Sea of Sorrow: Our Minamata Disease*, trans. Livia Monnet (1972; Kyoto: Yamaguchi Publishing, 1990). "The title . . . alludes to the contrast between the beauty of Minamata and the Shiranui Sea, and the agonizing suffering brought about by Minamata Disease, with the implication that the Shiranui Sea, once as beautiful as paradise, has become a source of crippling illness and death." Translator's introduction, vi. (Further references to this work, abbreviated *PSS*, will be included parenthetically in the text of chapter 6.)

10. The Kumamoto Nichinichi Cultural Prize, 1969; the Oya Sōichi Prize for Non-Fictional Literature, 1970; and the Republic of the Philippines' Ramón Magsaysay Prize, 1973. Ishimuri refused to accept any honors so long as the needs of Minamata Disease victims were ignored.

11. *Chisso* means "nitrogen" in Japanese; the company at first manufactured only nitrogenous fertilizers. It later expanded into a petrochemical concern, employing mercury as a catalyst in the manufacture of acetaldehyde used for plastics, drugs, perfumes, and photographic chemicals, among other products.

12. Comparable to a U.S. state legislature.

13. *Business Week*, 27 June 1977, 38.

CHAPTER SEVEN

1. Hedrick Smith, *The New Russians* (New York: Random House, 1990), 300.

2. Murray Feshbach and Alfred Friendly, Jr., *Ecocide in the USSR* (New York: Basic Books, 1992), 1, 93.

3. Maria Cherkasova et al., "Russia: Gasping for Breath, Choking on Waste, Dying Young," trans. Eliza Klose, *Washington Post*, 18 August 1991.

4. Maria Cherkasova, speech before the Institute of Soviet-American Relations (ISAR), Washington, D.C., 29 May 1990, *WorldWIDE News*, May–June 1990.

5. "Russia's population is shrinking by nearly 1 million, or 0.6 percent a year—the fastest decline ever recorded in an industrial society." Lester Brown, "Overview: A Year of Contrasts," in *Vital Signs, 1997: The Environmental Trends That Are*

Shaping Our Future, ed. Lester Brown (Washington, D.C.: Worldwatch Institute, and New York: W. W. Norton, 1997), 21.

6. Ludmilla Thorne, "Legacy of Chernobyl," *Baltimore Evening Sun*, 4 May 1994.

7. Vladimir Chernousenko, "We Shall Be Killed in Silent Ways," *Earth Island Journal*, winter 1995–96. (Chernousenko, a Ukrainian nuclear physicist, is himself dying of cancer caused by participating in the cleanup.) See also Thomas Orszag-Land, "Future Chernobyls Lurk along Russia's Northern Rim," *Christian Science Monitor*, 1 August 1996.

8. "When Soviet scientist Zhores Medvedev, living in Britain, wrote in 1976 that a tank containing 70 to 80 metric tons of radioactive waste had exploded in Russia's southern Urals in 1957, Western scientists vilified him in the press. . . . Not only did the Soviet government keep the accident secret from its people, but the United States government never revealed that the Central Intelligence Agency (CIA) knew the details as early as 1959." Melissa Akin, Brian Humphreys, and Lucian Kim, "On the Paper Trail of Cold War Secrets," *Christian Science Monitor*, 19 August 1996.

9. Maria Cherkasova, "Sounding the Alarm," *UNESCO Courier*, March 1992, 19.

10. Ibid., 22.

11. Marlise Simons, "Nowhere to Hide?" *St. Louis Post-Dispatch*, 11 April 1990.

12. Józef Niweliński, "Morphochemistry of Human Placentas in the Skawina Region" (in Polish with summary in English), *Folia Medica Cracoviensia* 23 (1981): 347–54.

13. Maria Gumińska, "Women in the Polish Ecological Club (PKE) Work for Environmental Protection," trans. Maria Gumińska and Malgorzata Niewiara, in *Proceedings of Seminar, "The Role of Women in Environmental Protection"* (Kraków: Polish Ecological Club, 1995), VI: 79.

14. A relatively new disease caused by exposure to flourine.

15. Maria Gumińska, "Biochemical Mechanisms of Environmental Fluoride Action," *Archiwum Ochrony Środowiska*, 3–4 (1990): 105–15.

16. Maria Gumińska and Andrzej Delorme, eds., *The Ecological Disaster of Kraków* (in Polish) (Kraków: Polish Ecological Club, 1990).

17. Gumińska says the Soviet Union had built the plant as a "gift to Poland." Most of the output, however, was exported to the USSR; the Soviets could, in effect, control production, even though the plant was not officially under their jurisdiction.

18. To this day, Gumińska notes, it is not clear whether the martial law edict originated with Soviet or Polish Communist leaders.

19. Gumińska, "Women in the Polish Ecological Club," 81.

CHAPTER EIGHT

1. Carol Browner, "Letter from the Administrator," in *The U.S. EPA's 25th Anniversary Report: 1970–1995* (EPA Web site).

2. Carol Browner, speech before the plenary session of the biennial convention of the National Audubon Society, Washington, D.C., 10 June 1996; taped by the

author. (Further references to this speech, abbreviated *AS*, will be included parenthetically in the text of chapter 8.)

3. Amy Kaslow, "Browner Dons Gloves for EPA," *Christian Science Monitor*, 8 August 1995.

4. Political action committees (PACs) of 212 special interests contributed $18 million to the election campaigns of U.S. representatives, $11.4 million of which went to 209 members of Congress who voted for the seventeen riders aimed at limiting the EPA's enforcement ability. Sierra Club, "Take the Money . . . and Run" (San Francisco: Sierra Club, 1996).

5. Ted Williams, "Silent Scourge," *Audubon*, January–February 1997, 28.

6. Alexandra Marks, "Perseverance on Clean Air Pays Dividends for Browner," *Christian Science Monitor*, 27 June 1997.

7. Jon Mitchell, "Jewels of the Sea: Fight for Survival in Fragile Ecosystems," *Christian Science Monitor*, 27 September 1996.

8. International Program of the Nature Conservancy, data sheet on coral reefs, March 1996.

9. Don Hinrichsen, "Reef Revival," *Amicus Journal*, summer 1996, 22.

10. Jon Luoma, "Reef Madness," *Audubon*, November–December 1996, 24.

11. International Program of the Nature Conservancy, data sheet, March 1996.

12. Jan Hartke, "Rain Forests of the Sea," *HSUS News*, summer 1996, 40.

13. This figure represents the natural increase—birth rate minus death rate; it does not factor in immigration and emigration numbers. Information obtained in telephone call to Population Action International, Washington, D.C., 1997.

14. Ben Wattenberg, host of the PBS program *Think Tank*, believes that, given the current rate of decline, population growth in China and India will sink below replacement rate within one or two decades. "Briefly" page, *Hinduism Today*, July 1997, 20.

15. See the section on Theo Colborn in chapter 4 for discussion of hormone disruptors.

16. The International Union for the Conservation of Nature (IUCN), an umbrella group of governmental and nongovernmental organizations, now calls itself the World Conservation Union. Since 1988 Pat Waak has served as an official delegate to the general assemblies that are held every four years by the IUCN.

17. Patricia Waak, "The Cairo Plan and IUCN: Looking Ahead," *Environmental Strategy*, June 1995, 18.

18. Patricia Waak, "Shaping a Sustainable Planet; The Role of Nongovernmental Organizations," *Colorado Journal of International Environmental Law and Policy* 6, no. 2 (summer 1995): 358–59.

19. Patricia Waak and Kenneth Strom, eds., *Sharing the Earth: Cross-Cultural Experiences in Population, Wildlife, and the Environment* (New York: National Audubon Society, 1992).

20. Opponents of family planning raise concerns about the government funding abortions. Since 1973, *no* U.S. foreign assistance moneys have been used for that purpose.

21. Waak, "Shaping a Sustainable Planet," 346.

22. "In 1996 the Walt Disney Co. spent millions on plans for a theme park in northern Virginia—only to find the idea scuttled by community opposition. Wal-Mart was prevented from opening stores in Massachusetts, Vermont, and Virginia by activist groups of maybe a dozen people each." Edmund M. Burke, "Forget the Government. It's the Community That Can Shut You Down," *Business Ethics*, May–June 1997, 11.

CHAPTER NINE

1. Gro Harlem Brundtland, chairman's foreword to *Our Common Future: The World Commission on Environment and Development* (New York: Oxford University Press, 1987), ix-x.
2. Ibid., xiii.
3. Gro Harlem Brundtland, "How to Secure Our Common Future," *Scientific American*, September 1989, 190.
4. Gro Harlem Brundtland, "Brundtland on Population," Rafael M. Salas Lecture at the United Nations, New York City, 28 September 1993 (Internet).
5. Hilary F. French, "Sustainable Development Aid Threatened," in *Vital Signs, 1997: Environmental Trends That Are Shaping Our Future*, ed. Lester Brown (Washington, D.C.: Worldwatch Institute, and New York: W. W. Norton, 1997), 108.
6. Charles J. Hanley, Associated Press, *America On Line* news, 27 June 1997.
7. Brundtland, "Brundtland on Population."
8. Hazel Henderson, *Paradigms in Progress: Life beyond Economics* (Indianapolis: Knowledge Systems, 1991), 9. (Further references to this work, abbreviated *PP*, will be included parenthetically in the text of chapter 9.)
9. Hazel Henderson, *Building a Win-Win World: Life beyond Global Economic Warfare* (San Francisco: Berrett-Koehler Publishers, 1996), 146. (Further references to this work, abbreviated *BW*, will be included parenthetically in the text of chapter 9.)
10. A University of Maryland ecological economist, Robert Costanza, calculates the dollar value of the planet's ecosystem services at an average of $33 trillion per year, versus global GNP of $18 trillion. Sharon Begley, "Butterflies Aren't Free," *Newsweek*, 26 May 1997, 73.
11. United Nations Development Programme, *UN Development Report, 1995* (New York: Oxford University Press, 1995), 97.
12. Henderson lists "obsolete sectors (unsustainable)" and "emerging sectors" side by side in a chart in *Building a Win-Win World*, 37.
13. Helena Norberg-Hodge, *Ancient Futures: Learning from Ladakh* (San Francisco: Sierra Club Books, 1992). (Further references to this work, abbreviated *AF*, will be included parenthetically in the text of chapter 9.)
14. In 1986 Norberg-Hodge and the Ladakh Project received the Right Livelihood Award.
15. See also David Korten, *When Corporations Rule the World* (West Hartford, Conn.: Kumarian Press, and San Francisco: Berrett-Koehler Publishers, 1995).

16. This group is dedicated to countering the threats posed to economic, social, cultural, environmental, and democratic rights by corporate economic globalization.

17. Vandana Shiva, *Monocultures of the Mind* (London: Zed Books, 1993), 135.

18. Vandana Shiva, *Biopiracy: The Plunder of Nature and Knowledge* (Boston: South End Press, 1997).

19. Vandana Shiva, "Biodiversity and Biopiracy," speech at the University of Washington, Seattle, for the benefit of the Edmonds Institute, 16 January 1996.

20. Ibid.

21. Vandana Shiva, *The Violence of the Green Revolution: Third World Agriculture, Ecology, and Politics* (London: Zed Books, 1991), 260.

22. Shiva, "Biodiversity and Biopiracy."

23. Shiva, *Monocultures of the Mind*, 120.

24. After the battles and starvation of the Gurkha wars of the early nineteenth century, huts in Himalayan communities were found empty of everything but the emaciated dead and full seed containers, which had remained untouched. Shiva, "Biodiversity and Biopiracy." And, "During the siege of Leningrad in World War II, scientists at the Vavilof Institute starved at their desks rather than eat their collection of seeds." Peter Tonge, "Interest Grows in Digging for Tomato's Family Roots," *Christian Science Monitor*, 24 March 1997.

25. Farmers who save and replant the product of a patented seed violate U.S. patent law. Hope Strand, "Patenting the Planet," *Multinational Monitor*, June 1994, 9–13.

26. Jeremy Seabrook, *Pioneers of Change: Experiments in Creating a Humane Society* (Philadelphia: New Society Publishers, 1993), 22.

27. Vandana Shiva, *Staying Alive: Women, Ecology, and Development* (London: Zed Books, 1988), 2, 4.

28. Ibid., 96.

29. Ibid., 152.

30. Michael Tobias, "World War 3: Population and the Biosphere," *Lapis*, no. 4 (1997): 25.

31. Shiva, *Monocultures of the Mind*, 102–3.

32. "Most plants absorb nitrogen as they grow, but beans and other legumes return nitrogen to the soil, reducing the need for chemical fertilizers." Joel Simon, "An Organic Coup in Cuba?" *Amicus Journal*, winter 1997, 35.

33. Shiva, *Staying Alive*, 139, 136–37.

34. Raising more than one plant species at the same time and place.

35. "Despite a tenfold increase in the use of chemical pesticides since World War II, the loss of food and fiber crops to insects has risen from seven to thirteen percent." Ted Williams, "Lethal Scourge," *Audubon*, January–February 1997, 28.

36. Shiva, *Monocultures of the Mind*, 7.

37. Shiva, *Violence of the Green Revolution*, 246.

38. "Vandana Shiva Talks to Judithe Bizot," *UNESCO Courier*, March 1992, 8.

39. Quoted in Steve Lerner, *Beyond the Earth Summit: Conversations with Advocates of Sustainable Development* (Bolinas, Calif.: Common Knowledge Press, 1992), 77.

40. The Women's Environment and Development Organization, founded by Bella Abzug.

41. Quoted in Lerner, *Beyond the Earth Summit,* 84.

42. "A year before the Earth Summit, Merck and Company, the world's largest phar-
maceuticals firm, and Costa Rica's National Institute of Biodiversity (INBio)
reached a precedent-setting bioprospecting agreement of the kind envisioned by
the treaty. Merck agreed to pay $1.35 million in support of INBio conservation
programs in exchange for access to the country's plants, microbes, and insects.
Should any Merck discovery make its way into a product, INBio will also collect
royalties." Hilary French, "When Foreign Investors Pay for Development,"
World Watch, May–June 1997, 9.

43. "Vandana Shiva Talks," 9.

44. Petra Kelly, *Thinking Green* (Berkeley, Calif.: Parallax Press, 1994), 61, 65. (Further references to this work, abbreviated *TG,* will be included parenthetically in
the text of chapter 9.)

45. Petra Kelly, *Fighting for Hope* (Boston: South End Press, 1984).

46. Petra Kelly, Gert Bastian, and Pat Aiello, eds., *The Anguish of Tibet* (Berkeley,
Calif.: Parallax Press, 1991).

47. As of 1997, the Green Party remained a strong influence, even though it held no
seats in the Bundestag.

48. In 1996 Arcata, California, became the first municipality in the United States to
elect a Green Party majority to its town council. Cathy Madison, "Arcata, CA:
Green Town in the Redwoods," *Utne Reader,* May–June 1997, 53.

49. But, early in 1997, the German government export credit agency—after secret
deliberations—offered financial loan guarantees to German makers of power-
generating equipment who were bidding for contracts on the Three Gorges Dam
project in China. This brought on angry reactions from the Green Party.

50. "An Unfitting End," *Time,* 2 November 1992, 14.

51. Hazel Wolf, "The McCarthy Period," speech to Rainier Beach High School,
Seattle, 24 May 1990 (Wolf's Web site.)

52. "Real People, Real Lives," *USA [Today] Weekend,* 30 August–1 September 1991.

53. According to a social research survey cosponsored by the Fetzer Institute and the
Institute of Noetic Sciences, a new cultural form is emerging, which the sociologist Paul H. Ray terms a "transmodern" or an "Integral Culture" that synthesizes
modernism and traditionalism. Leading the transformation is a group of forty-
four million adult Americans—a quarter of the adult population—who are leading-edge thinkers, or "Cultural Creatives." There are *50 percent* more women
than men in this group. Paul H. Ray, "The Rise of Integral Culture," *Noetic
Sciences Review,* spring 1996, 4.

CHAPTER TEN

1. Marjory Stoneman Douglas, with John Rothchild, *Marjory Stoneman Douglas:
Voice of the River* (Sarasota, Fla.: Pineapple Press, 1987), 190.

2. Marjory Stoneman Douglas, *The Everglades: River of Grass,* rev. ed. (Sarasota, Fla.:
Pineapple Press, 1987), 10.

3. Douglas, *Marjory Stoneman Douglas*, 191.

4. Douglas, *The Everglades*, 251, 286.

5. Douglas, *Marjory Stoneman Douglas*, 228.

6. "The Women's National Rivers and Harbours Congress in 1908 (USA) addressed water pollution and was instrumental in the creation of the National Conservation Congresses of 1909–1912. Although the Women's Congress lost their fight to ban the Hetch Hetchy dam in the Yosemite Valley (later to become a national park), their efforts resulted in the passing of the US National Parks Act in 1916. This transferred the care of national parks from the control of the National Forest Service, whose job was to manage the exploitation of forests as a natural resource, to the newly created National Parks Service for preservation." Bhaskar Nath, Luc Hens, and Dimitri Devuyst, eds., *Sustainable Development* (Brussels: VUB University Press, 1996), 65.

7. Douglas, *Marjory Stoneman Douglas*, 224.

8. Ibid., 230.

9. Warren Richey, "Reviving Florida's Fragile 'River of Grass,'" *Christian Science Monitor*, 3 September 1997.

10. Charles Osgood, "Seeds of Trouble," *CBS Sunday Morning*, transcript, 2 February 1997, 4.

11. Marjory Stoneman Douglas, "How You Can Protect the Environment," *GeoJourney*, October 1980, 13.

12. Douglas, *The Everglades*, 385.

13. Marjorie H. Carr, "The Ocklawaha River Wilderness," *Florida Naturalist*, August 1965, 1.

14. Marjorie Carr, "Opinion: A Golden Opportunity," *Monitor*, summer 1995, 2.

15. Marjorie Carr, "An Environmental Ethos for Florida," *Monitor*, March–April 1990, 2.

16. "In Western Europe, France is the only country still building nuclear plants, with three units in the pipeline." Nicholas Lenssen, "Nuclear Power Inches Up," in *Vital Signs, 1997: The Environmental Trends That Are Shaping Our Future*, ed. Lester Brown (Washington, D.C.: Worldwatch Institute, and New York: W. W. Norton, 1997), 48.

CHAPTER ELEVEN

1. Joseph Kastner, "Long Before Furs, It Was Feathers That Stirred Reformist Ire," *Smithsonian*, July 1994, 97.

2. Ibid., 100.

3. Ibid., 104.

4. Robert Cahn, *The Fight to Save Alaska* (New York: National Audubon Society, 1982), 4.

5. Margaret Murie, *Two in the Far North* (New York: Alfred A. Knopf, 1962), 153.

6. Ibid., 266.

7. *New York Times*, 20 June 1995, op-ed page.

8. Associated Press, "Go Slow in ANWR, Biologists Warn," *Anchorage Daily News*, 18 August 1995.

9. A study by the World Conservation Union indicates that nearly one-fourth of the planet's known mammal species are threatened with extinction, the most significant factor being destruction of habitat through human population growth and economic development. Curtis Runyan, "Mammals in Global Decline," *World Watch*, January–February 1997, 7.

10. According to Greenpeace, global warming has shrunk forty feet off the length of Bering Glacier in Alaska's Vitus Lake during the past century. Brad Knickerbocker, "Sudden New Heat on Global Warmth," *Christian Science Monitor*, 7 August 1997.

11. "WWF Report Indicates Arctic Species under Serious Threat from Global Warming," World Wildlife Fund, press release, Washington, D.C., 17 December 1996.

12. Sylvia Earle, *Sea Change: A Message of the Oceans* (New York: G. P. Putnam's Sons, 1995), 84–85.

13. Sylvia Earle, "Inner Resources," *Amicus Journal*, spring 1997, 56.

14. Earle, *Sea Change*, 172.

15. Ibid., xii–xiii.

16. Ibid., 325.

17. William D. Blair, Jr., *Katharine Ordway: The Lady Who Saved the Prairies* (Washington, D.C.: Nature Conservancy, 1989), 13.

18. Douglas H. Chadwick, "What Good Is a Prairie?" *Audubon*, November–December 1995, 36.

19. Ibid., 40.

20. Blair, *Katharine Ordway*, 41.

21. Ibid., 13.

22. Chadwick, "What Good Is a Prairie?" 43.

CHAPTER TWELVE

1. Bella Abzug, interview, 24 April 1997, Global Education Motivators Web site.

2. Bella Abzug, preface to *Women Making a Difference: An Action Guide to Women's Gains and Goals* (New York: Women's Environment and Development Organization, 1992).

3. Bella Abzug, "Women and the Environment," *Focus on Women*, bulletin, International Authors Series, U.N. Department of Public Affairs, April 1995.

4. Bella Abzug, Bradford Morse Memorial Gender and Development Lecture, Beijing, 5 September 1995, WEDO *News and Views* 8, no. 3–4 (December 1995).

AFTERWORD

1. For the "precautionary principle," see Bhaskar Nath, Luc Hens, and Dimitri Devuyst, eds., *Sustainable Development* (Brussels: VUB University Press, 1996), 33–36, 86.

2. Paul H. Ray, "The Rise of Integral Culture," *Noetic Sciences Review*, spring 1996, 4.

American Council of Learned Societies. *Dictionary of American Biography*. New York: Charles Scribner's Sons, 1963.

Anderson, Lorraine, ed. *Sisters of the Earth*. New York: Vintage Books, 1991.

Bari, Judi. *Timber Wars*. Monroe, Maine: Common Courage Press, 1994.

Behn, Sarala. "A Blueprint for Survival of the Hills." Supplement to *Himalaya: Man and Nature*. New Delhi: Himalaya Seva Sangh, 1980.

Blair, William D., Jr. *Katharine Ordway: The Lady Who Saved the Prairies*. Washington, D.C.: Nature Conservancy, 1989.

Boulding, Elise. *The Underside of History*. Newbury Park, Calif.: Sage Publications, 1992.

Brooks, Paul. *The House of Life: Rachel Carson at Work*. Boston: Houghton Mifflin, 1972.

Brown, Lester, ed. *State of the World, 1997*. Washington, D.C.: Worldwatch Institute; New York: W. W. Norton, 1997.

———. *Vital Signs, 1997: The Environmental Trends That Are Shaping Our Future*. Washington, D.C.: Worldwatch Institute, and New York: W. W. Norton, 1997.

———. *Vital Signs, 1996: The Environmental Trends That Are Shaping Our Future*. Washington, D.C.: Worldwatch Institute, and New York: W. W. Norton, 1996.

Browner, Carol. "Letter from the Administrator." In *The U.S. EPA's 25th Anniversary Report: 1970–1995*. Washington, D.C.: Environmental Protection Agency, 1995.

Brundtland, Gro Harlem. Chairman's Foreword. *Our Common Future: The World Commission on Environment and Development*. New York: Oxford University Press, 1987.

Cahn, Robert. *The Fight to Save Alaska*. New York: National Audubon Society, 1982.

Caldicott, Helen. *If You Love This Planet*. New York: W. W. Norton, 1992.

Carson, Rachel. *Silent Spring*. Boston and New York: Houghton Mifflin, 1962.

Clarke, Robert. *Ellen Swallow, The Woman Who Founded Ecology*. Chicago: Follette Publishing, 1973.

Colborn, Theo, Dianne Dumanoski, and John Peterson Myers. *Our Stolen Future: Are We Threatening Our Fertility, Intelligence, and Survival?—A Scientific Detective Story*. New York: Dutton, 1996.

Congress and the Nation—1945–1964: A Review of Government and Politics. Ed. Thomas N. Schroth. Washington, D.C.: Congressional Quarterly Service, 1965.

Dai Qing, ed. *Yangtze! Yangtze!* Trans. Nancy Liu et al. Toronto and London: Probe International and Earthscan, 1994.

Douglas, Marjory Stoneman. *The Everglades: River of Grass*. Rev. ed. Sarasota, Fla.: Pineapple Press, 1987.

Douglas, Marjory Stoneman, with John Rothchild. *Marjory Stoneman Douglas: Voice of the River*. Sarasota, Fla.: Pineapple Press, 1987.

Earle, Sylvia. *Sea Change: A Message of the Oceans*. New York: G. P. Putnam's Sons, 1995.

Ehrlich, Paul R., and Anne H. Ehrlich. *Healing the Planet*. New York: Addison-Wesley Publishing, 1995.

Feshbach, Murray, and Alfred Friendly, Jr. *Ecocide in the USSR*. New York: Basic Books, 1992.

Fuller, Kathryn. Foreword to *World Wildlife Fund Annual Report, 1996*. Washington, D.C.: World Wildlife Fund, 1996.

Garland, Anne Witte. *Women Activists—Challenging the Abuse of Power*. New York: Feminist Press, 1988.

Gearhart, Sally Miller. "The Future—If There Is One—Is Female." In *Reweaving the Web of Life*, edited by Pam McAllister. Philadelphia: New Society Publishers, 1982.

Gibbs, Lois, as told to Murray Levine. *Love Canal—My Story*. Albany: State University of New York Press, 1982.

Gilpin, Laura. *The Rio Grande, River of Destiny: An Interpretation of the River, the Land, and the People*. New York: Duell, Sloan and Pearce, 1949.

Goldman Environmental Foundation. *Goldman Environmental Prize: The First Five Years*. San Francisco: Goldman Environmental Foundation, 1995.

Gore, Al. *Earth in the Balance: Ecology and the Human Spirit*. New York: Penguin Books, 1993.

Graham, Frank. *Since "Silent Spring."* Boston: Houghton Mifflin, 1972.

Guide to American Law, Everybody's Legal Encyclopedia. Ed. Harold W. Chase. 12 vols. New York: West Publishing, 1984.

Gumińska, Maria. "Women in the Polish Ecological Club (PKE) Work for Environmental Protection." In *Proceedings of Seminar, "The Role of Women in Environmental Protection."* Kraków: Polish Ecological Club, 1995.

Gumińska, Maria, and Andrzej Delorme, eds. *The Ecological Disaster of Kraków* (in Polish). Kraków: Polish Ecological Club, 1990.

Haley, Delphine. *Dorothy Stimson Bullitt: An Uncommon Life*. Seattle: Sasquatch Books, 1995.

Hawken, Paul. *The Ecology of Commerce.* New York: HarperBusiness Division of HarperCollins Publishers, 1993.

Henderson, Hazel. *Building a Win-Win World: Life beyond Global Economic Warfare.* San Francisco: Berrett-Koehler Publishers, 1996.

———. *Creating Alternative Futures: The End of Economics.* New York: G. P. Putnam's Sons, 1978.

———. *Paradigms in Progress: Life beyond Economics.* Indianapolis: Knowledge Systems, 1991.

———. *The Politics of the Solar Age: Alternatives to Economics.* Indianapolis: Knowledge Systems, 1981; rev. ed. 1988.

Henderson, Hazel, ed., with Harlan Cleveland and Inge Kaul. *The United Nations: Policy and Financing Alternatives.* Washington, D.C.: Global Commission to Fund the United Nations, 1995.

Hoy, Suellen M. "Municipal Housekeeping: The Role of Women in Improving Urban Sanitation Practices, 1880–1917." In *Pollution and Reform in American Cities—1870–1930,* edited by Martin Melosi. Austin: University of Texas Press, 1980.

Hozeski, Bruce. *Hildegard of Bingen: The Book of the Rewards of Life.* New York: Garland Publishers, 1993.

Hunt, Carolyn. *The Life of Ellen H. Richards.* Boston: Whitcomb and Barrows, 1912.

Ishimuri, Michiko. *Paradise in the Sea of Sorrow: Our Minamata Disease.* Translated by Livia Monnet. Kyoto: Yamaguchi Publishing, 1990. First published in Japanese in 1972.

Jacobsen, Judith E. "Population, Consumption, and Environmental Degradation: Problems and Solutions." *Colorado Journal of International Environmental Law and Policy* 6, no. 2 (summer 1995): 255–72.

James, Edward T., ed. *Notable American Women—1607 to 1950.* Cambridge, Mass.: Belknap Press of Harvard University Press, 1971.

Janeway, Elizabeth. *Between Myth and Morning—Women Awakening.* New York: William Morrow, 1974.

Kelly, Petra. *Fighting for Hope.* Boston: South End Press, 1984.

———. *Thinking Green.* Berkeley, Calif.: Parallax Press, 1994.

Kelly, Petra, Gert Bastian, and Pat Aiello, eds. *The Anguish of Tibet.* Berkeley, Calif.: Parallax Press, 1991.

Korten, David. *When Corporations Rule the World.* West Hartford, Conn.: Kumarian Press, and San Francisco: Berrett-Koehler Publishers, 1995.

LaBastille, Anne. *Women and Wilderness.* San Francisco: Sierra Club Books, 1980.

Lerner, Steve. *Beyond the Earth Summit: Conversations with Advocates of Sustainable Development.* Bolinas, Calif.: Common Knowledge Press, 1992.

Mathews, Jessica Tuchman, ed. *Preserving the Global Environment—The Challenge of Shared Leadership.* New York: W. W. Norton, 1991.

McHenry, Robert, ed. *Liberty's Women.* Springfield, Mass.: G and C Merriam, 1980.

Meadows, Donella. *Global Citizen.* Washington, D.C.: Island Press, 1991.

Means, Russell, with Marvin J. Wolf. *Where White Men Fear to Tread: The Autobiography of Russell Means.* New York: St. Martin's Press, 1995.

Melosi, Martin, ed. *Pollution and Reform in American Cities, 1870–1930.* Austin: University of Texas Press, 1980.

Merchant, Carolyn. *The Death of Nature: Women, Ecology, and the Scientific Revolution.* San Francisco: Harper, 1989.

Montagu, Ashley. *The Natural Superiority of Women.* New York: Macmillan, 1953.

Montgomery, Sy. *Walking with the Great Apes.* Boston: Houghton Mifflin, 1991.

Murie, Margaret. *Two in the Far North.* New York: Alfred A. Knopf, 1962.

Nath, Bhaskar, Luc Hens, and Dimitri Devuyst, eds. *Sustainable Development.* Brussels: VUB University Press, 1996.

Nichols, William Ripley. "The Present Condition of Certain Rivers of Massachusetts Together with Considerations Touching the Water Supply of Towns." *Fifth Annual Report.* Massachusetts State Board of Health, January 1874, Boston.

Norberg-Hodge, Helena. *Ancient Futures: Learning from Ladakh.* San Francisco: Sierra Club Books, 1991.

O'Neill, Lois Decker, ed. *The Women's Book of World Records and Achievements.* Garden City, N.Y.: Anchor Books, 1979.

Plutarch. *The Parallel Lives.* Sertorius and Eumenes. Loeb Classical Library, vol. 8, sec. 16. New York: G. P. Putnam's Sons, 1919.

Reisner, Marc. *Cadillac Desert: The American West and Its Disappearing Water.* New York: Viking Penguin, 1986.

Richards, Ellen Swallow. *Air, Water, and Food from a Sanitary Standpoint.* New York: John Wiley and Sons, 1901.

———. *Conservation by Sanitation.* New York: John Wiley and Sons, 1911.

———. *Food Materials and Their Adulterations.* Boston: Estes and Lauriat, 1886.

———. *Sanitation in Daily Life.* Boston: Whitcomb and Barrows, 1907.

Seabrook, Jeremy. *Pioneers of Change: Experiments in Creating a Humane Society.* Philadelphia: New Society Publishers, 1993.

Shiva, Vandana. *Biopiracy: The Plunder of Nature and Knowledge.* Boston: South End Press, 1997.

———. *Monocultures of the Mind.* London: Zed Books, 1993.

———. *Staying Alive: Women, Ecology, and Development.* London: Zed Books, 1988.

———. *The Violence of the Green Revolution: Third World Agriculture, Ecology, and Politics.* London: Zed Books, 1991.

Smith, Hedrick. *The New Russians.* New York: Random House, 1990.

Smith, W. Eugene, and Eileen M. Smith. *Minamata.* New York: Holt, Rinehart and Winston, 1975.

Stern, Madeleine B. *We the Women: Career Firsts of Nineteenth-Century America.* Omaha: University of Nebraska Press, 1994.

Terkel, Studs. *Coming of Age: The Story of Our Century by Those Who've Lived It*. New York: New Press, 1995.

United Nations Development Programme. *UN Development Report, 1995*. New York: Oxford University Press, 1995.

U.S. Department of Commerce, Bureau of the Census. *Historical Statistics of the U.S., Colonial Times to 1970*, Series B 201–13. Washington, D.C., 1975.

Waak, Patricia. "The Cairo Plan and IUCN: Looking Ahead." *Environmental Strategy*, newsletter of the Commission on Environmental Strategy and Planning, IUCN (World Conservation Union). Gland, Switzerland, June 1995.

———. "Shaping a Sustainable Planet: The Role of Nongovernmental Organizations." *Colorado Journal of International Environmental Law and Policy* 6, no. 2 (summer 1995): 345–62.

Waak, Patricia, and Kenneth Strom, eds. *Sharing the Earth: Cross-Cultural Experiences in Population, Wildlife, and the Environment*. New York: National Audubon Society, 1992.

Wallace, Aubrey. *Eco-Heroes: Twelve Tales of Environmental Victory*. San Francisco: Mercury House, 1993.

Weber, Thomas. *Hugging the Trees: The Story of the Chipko Movement*. New York: Penguin Press, 1989.

Who Was Who in American History—Science and Technology. Chicago: Marquis Who's Who, 1976.

Wilson, Edward O. *In Search of Nature*. Washington, D.C.: Island Press, 1996.

Yost, Edna. *American Women of Science*. Philadelphia: J. B. Lippincott, 1943.